ISBN 978-1-330-23080-0
PIBN 10058704

1 MONTH OF
FREE
READING

at

www.ForgottenBooks.com

By purchasing this book you are eligible for one month membership to ForgottenBooks.com, giving you unlimited access to our entire collection of over 700,000 titles via our web site and mobile apps.

To claim your free month visit:

www.forgottenbooks.com/free58704

English
Français
Deutsche
Italiano
Español
Português

www.forgottenbooks.com

Mythology Photography **Fiction**
Fishing Christianity **Art** Cooking
Essays Buddhism Freemasonry
Medicine **Biology** Music **Ancient**
Egypt Evolution Carpentry Physics
Dance Geology **Mathematics** Fitness
Shakespeare **Folklore** Yoga Marketing
Confidence Immortality Biographies
Poetry **Psychology** Witchcraft
Electronics Chemistry History **Law**
Accounting **Philosophy** Anthropology
Alchemy Drama Quantum Mechanics
Atheism Sexual Health **Ancient History**
Entrepreneurship Languages Sport
Paleontology Needlework Islam
Metaphysics Investment Archaeology
Parenting Statistics Criminology
Motivational

THE

FORT WILLIAM HENRY HOTEL

Lake George, Warren Co., N. Y.

T. ROESSLE & SON, Proprietors.

ALSO PROPRIETORS OF

THE ARLINGTON,	-	-	*Washington, D. C.*
THE DELAVAN,	-	-	*Albany, N. Y.*

SEASON, FROM JUNE 1 TO OCTOBER 1.

During the past winter the house has been thoroughly overhauled and several important additions have been made, prominent among which may be mentioned, a new dining-room, giving a facility for dining comfortably one thousand guests. Through cars are now run daily from Grand Central Depot to Fort William Henry Hotel without change.

The American Fall—Whitney.

OUR AMERICAN RESORTS.

For Health, Pleasure, and Recreation.

WHERE TO GO AND HOW TO GET THERE.

CONCERNING THE SUMMER AND WINTER RESORTS IN THE UNITED
STATES AND CANADA, FROM SEA-SHORE TO MOUNTAIN AND GLEN,
AND FROM THE GROVES OF SUNNY FLORIDA TO THE
LAKES AND PARKS OF THE NORTH AND WEST,
WITH THE CHARACTERISTICS OF EACH,
THEIR LEADING ATTRACTIONS, HO-
TEL ACCOMMODATIONS, LO-
CATION, MEANS OF AP-
PROACH, RAILWAY
FARES, ETC.

With nearly One Hundred and Fifty Illustrations.

EDITED BY LOUIS M. BABCOCK.

FIRST EDITION.

WASHINGTON, D. C.:

NATIONAL NEWS BUREAU.

1883.

PREFACE.

THE work of planning, gathering materials for and prepaiing this edition of OUR AMERICAN RESORTS was not begun until late in the season, or about the time it should have been ready for the press. It has, therefore, been hurriedly done. Of its shortcomings and imperfections the editor is quite as well aware as will be its severest critics. The difficulties encountered in the preparation of such a work are multitudinous, not the least of which is the tardiness and lack of enterprise among those whose interest it is to have information concerning themselves widely disseminated. For this unbusiness-like backwardness, the many worthless and dishonest publications through which they have been annoyed and defrauded are, no doubt, largely responsible. But this, added to the great difficulty in reaching proprietors of resorts at a season when they are not open, has rendered any approach to completeness in the present issue impossible. Still, the work as it is compares favorably with any of its class heretofore published, and ultimately it shall be vastly superior to them all.

Nothing like a Directory has been attempted. A work of that character sufficiently complete to embrace all the nooks and hamlets that consider themselves resorts would require more than a thousand pages of closely-condensed matter, involve interminable labor, and possess no special interest when issued. In these pages an effort has been made to present in readable form—with numerous illustrations for embellishment and aid—correct impressions of the important resorts and natural wonders within the borders of our own land, including something concerning the characteristics and climates of the localities in which they are situated. The work therefore, without being strictly either a Guide, Gazetteer or Handbook, contains much information pertaining to the realm of travel for health or sight-seeing.

On the first of May next a thoroughly revised edition will be issued, containing all that is worth gathering to complete the work, with over one hundred additional illustrations, many of them from new and original sketches, engraved by the best artists. The aim will be to embrace not only all places worthy of note, but to incorporate the fullest details concerning them, their accommodations, and the various routes of travel. A table of railway and steamboat fares between resorts named and the principal cities will also be added. Meanwhile correspondence is solicited with all who may be interested.

WASHINGTON, D. C., June 1st, 1883.

(iv)

CONTENTS.

(v)

THE PROLOGUE.

S UMMER recreation has come to be a recognized necessity. Rest, change, and relaxation are natural requirements of the human system, and especially of the dwellers in cities, whose lives partake so much of the artificial, and who are so far removed, as it were, from nature and the influences of outdoor freedom. Hercules could not, at first, conquer Antæus—the son of earth and sea—because each time the giant was thrown he gained new strength from mother earth. The parallel is easily drawn. As the human mind and body need sleep, as they should have one day in seven for rest, so do they require each year a period during which they may escape from the toil and vexations of business, the wear, and grind, and the routine of usual avocations, and gain new vigor by simple contact with nature, breathing the air, using the diet, seeing the sights, and hearing the sounds of the country. Cowper expresses a homely truth in his lines :

> " God made the country and man made the town ;
> What wonder, then, that health and virtue, gifts
> That can alone make sweet the bitter draught
> That life holds out to all, should most abound
> And least be threatened in the fields and groves ?"

In this age the feverish excitement of speculation, the sharp competition in business, and the close application and incessant activity of professional and business men, with high rates of living and social dissipations, all combine to break down health in our cities and render a season of recuperation doubly necessary. It is not, therefore, for mere idle pleasure and sight-seeing alone that such a large proportion of city people now annually spend a part of the heated term away from home—in the mountains, at the springs, or the sea-shore,—and that the number is increasing year by year. And at this season, when the first warm days foreshadow the warmer ones to come, the preparations for vacation begin—the casting about for a place to go occupies attention. It is the time when

> " The bleating lambs, of tender age,
> · Frisk gayly o'er the lawn.
> The sweatful farmer smites his mules
> And ploughs the growing corn.
> The city cousins pack their trunks
> And coo their softest coo,
> Dear Uncle—We'll be down next week
> And bring the children, too."

It is a good thing that in this world of ours a means of supply is provided for every real need; that as the seasons roll round with their ceaseless changes the genius of man is constantly devising ways of meeting and filling the requirements of the day and hour, thus making not only the waste places of the earth but of men's souls "blossom as the rose." With the increase of demand there is an increase of inducements, and every year new attractions are developed, new beauties and new wonders discovered. There are summer resorts and summer resorts; places where the curious and vain may see and be seen, where nature in its loveliest and grandest aspects may be studied, or where tired humanity may refresh itself according to its bent. The great variety is only equalled by the vastly differing tastes and requirements. And as inclinations diverge one year, so will the same individuals recognize in themselves changed conditions and needs for the next. In this, as in all things, variety is the spice of life.

> "Of all the passions that possess mankind,
> The love of novelty rules most the mind;
> In search of this, from realm to realm we roam,
> Our fleets come fraught with ev'ry folly home."

Yet changing and changeful mankind, with many pleasures to choose, can sometimes find delight in none. Among those who have naught to seek but enjoyment, that is not infrequently found the most tiresome of all occupations, for satiety is a stubborn disease. Often the things which were erstwhile our pleasure to-day pall upon the taste; and a quotation from one of Pope's moral essays well depicts the humor of many who seek places of resort:

> "Papillia, wedded to her amorous spark,
> Sighs for the shades—'how charming is a park!'
> A park is purchased, but the fair he sees
> All bathed in tears—O, odious, odious trees."

But withal, summer resorts are a blessing to the race, and sick or well, rich or poor, all derive increase of years with increase of happiness from the days or weeks spent in rational recreation. In the succeeding pages are described many places of real interest, any of which will amply repay a visit, either for health or pleasure.

The City of Washington.

" Sun of the moral world! effulgent source
Of man's best wisdom and his steadiest force,
Soul-searching Freedom! here assume thy stand,
And radiate hence to every distant land."

T is certainly appropriate that a work on AMERICAN RESORTS emanating from the National Capital, should begin with something concerning the attractions of a city in which all Americans are interested, and to which thousands of tourists and sight-seers journey every month in the year. Washington is sought and visited by people from every section, at all seasons, not alone because it is the nation's capital, but also for the reason that it is the most beautiful and attractive of American cities, its climate the most salubrious, and its surroundings the most interesting. In spring and summer people come to see the beauties of the city, and to visit the places of public interest of which they have heard. In winter people of wealth come to enjoy the comforts of a mild climate and to

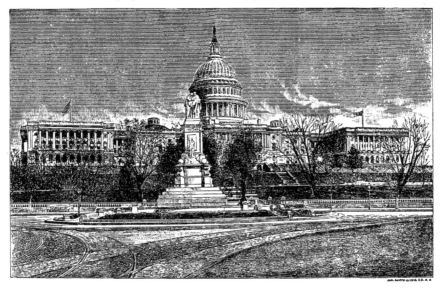

The Capitol.

participate in the round of semi-official social pleasures. The artist, the philosopher and the scholar find here a congenial workshop, and rich stores of the choicest fruits culled from nature, art and literature. So, to this city, already great and beautiful, but destined probably to be greater and more beautiful than Rome in its prime, come all the currents of the national life, a tide of vast magnitude, which increases in volume as the country grows in population and the attractions of the Capital multiply in number and variety. It was said of Rome that "as the streams lose themselves in the ocean, so the history of the peoples once distributed along the Mediterranean shores is absorbed in that of the great mistress of the world." Of Washington it may be said that it is rapidly becoming a storehouse of the products of the genius of all mankind. Our seat of government, apart from its political attractions, contains, even now, so much that is of interest in architecture and antiquities, such art collections and such treasuries of knowledge and invention in its museums and its Patent Office, as to compete almost on even terms with the great centres of commerce all combined. The actual population of Washington is not above two hundred thousand; but, like the human heart which it typifies, all the blood of the country, sooner or later, runs through it, and everybody is at one time or another a resident. The ebb and flow of transient visitors and temporary inhabitants is so enormous that railways alone can give prompt ingress and egress to the tide, and these railways, by the very facilities they furnish, but provoke a still greater volume of travel. Do you want to find a particular man on the street? Stand where you are and he will pass by after awhile. So, if you want to see anybody, you have only to go to Washington and wait a day or two; he will be sure to turn up. It is worth your while to visit the city, if only to be surprised by the sudden appearance of the very last person in the world that you ever expected to see.

The Smithsonian Institute.

Washington is located on the east bank of the Potomac, at the head of navigation, 295 miles from the ocean, where the river runs from the northwest to the southeast and expands to the width of over a mile. It is situated upon and surrounded by high bluffs and hills on the Maryland side, while on the opposite side are Arlington Heights and Fort Whipple. The District of Columbia was selected as the site of the National Capital after much consideration by Congress, extending over the period from October, 1783, to July, 1790; and on the 16th day of the latter month the act entitled "An act establishing the temporary and permanent seat of government of the United States" was passed by a vote of 32 to 29. Many and weighty were the reasons urged for the selection made, not the least among which was the deference and respect which would thus be paid to the wishes of General Washington, who from the first strongly advocated this point; his attention, it is said, having been fixed upon its advantages when a youthful surveyor of the country round. That he builded better than he knew is evidenced in the fact that the Washington of to-day eclipses the most sanguine expectations of its founder, and in beauty far surpasses the capitals of other nations. The design of L'Enfant who planned the city, although derided for seventy years, has through the genius and energy of Governor Shepherd been made to develop into a model of convenience and sightliness. Though many changes have been made since Webster-denominated Washington "a city of magnificent distances," its broad streets and numerous reservations are still suggestive of abundant breathing-room. These broad, well-paved and cleanly-swept streets, interspersed with parks, squares and fountains, are laid out in parallel lines from east to west and north to south, while the avenues radiating from the Capitol and Executive Mansion intersect them at various points, forming circles, triangles, and oblongs, all of which are beautifully adorned with trees, shrubbery and flowers. There is hardly a street or avenue but adown its vista some allurement is displayed; this one reaches far away through the green of maple and linden and the blue of distance across the long bridge to the hills of Virginia; that one ends in the lovely grounds of the Agricultural Department or the Smithsonian Institute, while Pennsylvania Avenue, like the kaleidoscope, presents new scenes at every turn. Almost every foot of its length and breadth is replete with historical incidents. Here ruler and ruled jostle each other; cabinet officers and senators and representatives are not distinguished above the common mass, and over this smooth roadway noiselessly rolls the liveried equipage of foreign ambassadors, side by side with the humble vehicle of the private citizen. Sooner or later all the famous of our own

White House—East View.

and other countries are tolerably sure to meet and pass upon this grand highway. Washington in his yellow chariot, drawn by six white horses, has driven over it; Hamilton, Lafayette, Clay, Webster, and all the gods of the Republic have trodden it. Five hundred thousand Federal soldiers marched up this avenue in review before the President and the Generals of the Army, shouting their songs of gladness, in 1865, when grim-visaged war had given way to white-winged peace. Those who remember the Washington of 1861-5, upon returning now for the first time, can scarcely realize that the Washington of to-day is the same city. Sections which were then outlying swamps, and others that were the abode of wretchedness and squalor, have been transformed into something like fairy land. Grand and imposing public buildings have been erected; squares and parks have been laid out and improved and beautified, and palatial private residences, with surrounding adornments, have risen up in every quarter.

Naturally the public buildings first attract the visitor's eye. The Capitol, the central figure, first awes, then allures by its imposing outline and proportions. It is not only the finest building in America, but in the world. Standing in all its magnificence upon one of the city's highest hills, the great white dome, surmounted by the colossal statue of the Goddess of Liberty, rises over the immense pile of granite like an imperishable signal of freedom for the oppressed of all the earth. At its base the greensward, velvety lawns and embowering trees betoken the shelter and repose found in the shadow of its ægis. The east façade or front of the building looks out over East Capitol Park and the plain of Capitol Hill, with the azure hills of "My Maryland" for a background, the west overlooking the business part of the city and the Potomac, commanding a view pronounced by the great traveller Humboldt one of the most beautiful his eyes had ever beheld. The Capitol is not only interesting and pleasing on account of the beauty of its exterior and surroundings, but it contains within some of the richest treasures of the nation.

Lee Mansion at Arlington.

First, there is the Congressional library, one of the largest and most valuable in the world; the Senate Chamber, Hall of Representatives, Gallery of Statuary, and the Rotunda filled with paintings by our greatest masters. A continuous park about two miles in length extends from the Capitol westward to the Potomac, within which are the Botanical Gardens and Greenhouses, the Smithsonian Institution, and New National Museum, containing a vast collection of natural curiosities and works of art, the Department of Agriculture, the Washington Monument, and the Bureau of Engraving and Printing, where all our bank notes and government stamps are made. The west end of this park sweeps round to the north, taking in the Treasury Building, the Executive Mansion, known as the White House, and that elegant new structure, a model of modern architecture, occupied by the State, War and Navy Departments. Besides these there are in other sections of the city, the Interior Department building, containing the bureaus of Indian Affairs, Public Lands, and the Patent Office, with the thousands of models stored there as monuments to the genius of American inventors, the Post-Office Department, the Medical Museum, the Naval Observatory, the Navy Yard, and many other points each entitled to a day's inspection. To minutely mention all the public buildings and institutions is not the province of this work. The visitor will find comprehensive guide-books easily obtained if needed, and will learn in a few days' sojourn many interesting facts and details not readily committed to paper.

The tourist to Washington usually comes with the notion that the Public Buildings, the President and Congress comprise about all there is worth seeing in the Capital. But, while these are among its chief attractions, there are many others equally calculated to interest and delight. Lincoln Park, with the colossal statue of the "Martyred President" striking the chains from the limbs of the slave, Lafayette Park, with Mills' Equestrian Statue of Jackson, the most wonderful artistic work of its class, Washington Circle, with its statue of him who was "first in war, first in peace, and first in the hearts of his countrymen," with various other parks and circles wherein are statues of Scott, Thomas, McPherson, Farragut and other national heroes present a study of interest. The Corcoran Art Gallery, the personal gift of the great philanthropist whose name it bears, is known throughout the land, and in its spacious halls are gathered some of the finest gems that have made immortal the artists of this and past generations. Besides all these interesting features within the city, there are other attractions no less important in its surroundings. The drives about Washington are unsurpassed, affording views of nature's wildest freaks as well as the cultivated splendors of romantic and æsthetic taste. Two hundred old forts crumbling to decay on the hills round about are eloquent reminders of our late civil war; Arlington, with its memories of Washington, Custis and Lee, and its thousands of mounds above the dead who died for country ; the Soldiers' Home, with its eight or nine hundred acres of park, its unrivalled drives stretching over hill and vale, and its wilderness of flowers and forest trees; Kalorama Heights, overlooking the city and affording a view southward as far as the eye can

Soldiers' Home.

reach, are each worthy of more than passing notice, and afford hours of pleasant, satisfying exploration. The scenery along the Potomac to Great Falls, 16 miles above Georgetown, is marvellous in its romantic beauty. A trip past the Georgetown Heights, over the conduit road, and past Cabin, John's Bridge, the longest single span in the world, is one of the many pleasant rides. The Great Falls themselves and their surroundings comprise a scene scarcely equalled anywhere for romantic beauty and ruggedness. Another of the delightful drives about the Capital is up Rock Creek, the stream which separates West Washington from the city proper. The country around this creek, though bordering upon the city and almost entering its very gates, to-day remains in the perfection of wildness and natural beauty. It is indeed an enchanting spot, replete with inviting retreats, leafy bowers and rippling waters.

" Nature was here so lavish of her store
That she bestowed until she had no more."

The garden spot of Washington, literally, is the portion south of Pennsylvania avenue, including what is termed the Mall and the White lot. In this area of several hundred acres are some beautiful drives, around the Botanical Garden, in the Smithsonian and Agricultural grounds, and around the other squares included. The report of the Parking Commission shows that there are to-day nearly one hundred and twenty miles of trees in the city of

Washington, of which about one-half are maples. The remainder includes poplars, box elders, elms, lindens, buttonwoods, willows and firs. These include only the work of the Parking Commission and represent the fruits of about ten years' labor. The young trees thrive well and give promise of making the city more and more attractive as yearly they increase in size. The stately giants in the public parks, also numbered by thousands, are not included in the count of nature's ornaments under the charge of this Commission.

Few visitors to Washington leave it without taking a trip to Mount Vernon, the former home and present resting-place of the "Father of his Country." It is an exceedingly pleasant excursion down the Potomac past Fort Washington and Fort Foote, and the scenery along the river is full of picturesque interest. The steamer Corcoran which makes this trip daily, leaving at 10 A. M., and returning at 4 P.M., is one of the finest vessels on the river, and is under the command of Captain L. L. Blake, well known as an experienced and genial officer. The tolling of the steamer's bell announces the approach to the tomb of Washington, in accordance with the

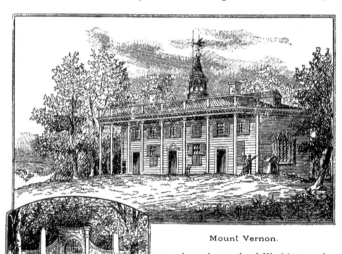

Mount Vernon.

custom among all steam vessels while passing Mount Vernon. Once in the grounds time passes so rapidly while wandering through the groves and gardens of this beautiful old homestead, standing high upon the bluff overlooking the river, and there is so much interest in looking through the quaint old rooms of the mansion, that the day seems all too short. It is not the object of this article to picture all the visitor may see in and around the National Capital, but the aim has been to direct attention to some of the most attractive features, and to give a reason for the faith that is in us—to tell how the tourist may be repaid for coming here. Washington is reached from the south by the Virginia Midland and the Richmond, Fredericksburg and Potomac Railways. From New York, Philadelphia and the East or West take the Pennsylvania Railroad. From the Southwest the Chesapeake and Ohio is the best route. The Baltimore and Ohio road also runs into Washington, and during the summer the Potomac Steamboat Company run a daily line to Norfolk connecting with steamers for New York, Boston and other points on the Atlantic coast. The hotel accommodations of the city are excellent and ample. The Arlington, Willard's, the Riggs, the Ebbitt, and other large first-class houses are unsurpassed in all their appointments. No city in the Union is better equipped for entertaining large numbers in comfort and luxury.

Virginia Resorts.

HE Old Dominion embraces within her borders such an extensive variety of health-restoring, pleasure-giving resorts as to be well entitled to a special classification of her own. Traversed by two ranges of mountains and innumerable winding rivers, the State abounds in beautiful valleys and incomparable landscapes, the magnificence and grandeur of which are not fully appreciated by many of her own people. One of the finest views the writer ever beheld may be seen from the summit of the Blue Ridge mountains near Snicker's Gap, looking down into the famous Loudon valley ; but there are a hundred others from various points in both the Blue Ridge and the Alleghanies nearly equal to it. Not only has Virginia her full share of mountains, high rugged crags and rocky slopes, but beautiful cascades, wonderful caves, and romantic glens are encountered in various portions of her domain. In the diversity, surprising character, and interesting features of the study she affords to lovers of the marvellous and picturesque, few localities can sustain a claim to superiority over the old State, which, in the early days of our national history, won distinction as "the mother of presidents."

New River at Nuttallburg, C. and O. Railway.

The mountains of Virginia do not point bold stony summits above the clouds far beyond vegetation and timber line, like the lofty peaks of the Rockies in Colorado or the Sierras in California, but they are high enough to be grand, while still retaining the charm and beauty of verdure. The climate, in general, is that of the temperate zone, and the mountain region is exceedingly healthy. The thermometer seldom rises higher than 85° in the hottest days, and the nights are always cool. Occupying a middle ground between the rigorous climate of the North and the enervating heat of the extreme South, Virginia is geographically one of the choicest sections of the country. The principal rivers of the State are the Potomac, the Greenbrier, the Rappahannock, the Shenandoah, the James, of which the Chickahominy is a tributary, the York, the New, and the Roanoke. Nearly all these are navigable to a considerable distance toward the mountains in which they generally take their rise, and their banks are highly picturesque. Beyond the limits of navigation the wildness of mountain streams obtain, and numerous waterfalls

Griffiths' Knob, C. and O. Railway.

of striking beauty are to be seen. Of these, the falls of the James River, and the New River Falls, seven miles from Hinton, on the Chesapeake and Ohio Railway, are perhaps the most noted. The cataract of the latter is but twenty-four feet high, but the width of the river and extent of the rapids make a scene of unusual wonder. Next to her mountains, the Mineral Springs of Virginia are her chief attraction. They are many in number and extensively varied in character. Some are famous throughout the country, the Old White Sulphur, for instance, having been a noted fashionable resort and political rendezvous for years before the Civil War. Several of the most prominent are situated on the line of the Chesapeake and Ohio Railway. The tourist over this line will enjoy one of the most picturesque journeys to be found on this continent. The scenery is not, perhaps, as sublime and startling as that along the narrow gauge roads of Colorado, but it is wild and abrupt, with all the softening tints of a fine painting. Along the Norfolk and Western and Shenandoah Valley roads there are also glimpses of such landscapes as few sections of country afford.

> " 'Tis beauty truly lent, whose red and white
> Nature's own sweet cunning hand laid on."

Every one who has ever crossed the Alps into Italy remembers the zigzags from which he looks down on the valley he is reaching, but without exaggeration it may be said that all the alternations of dark tunnel and picturesque valley of that famous little road could be sub-tracted from the Chesapeake and Ohio line without being missed. All travellers by the famous Pennsylvania Railroad remember that attractive piece of fancy engineering known as Horseshoe Bend, and nobody has gone to California without treasuring a recollection of the rounding of Cape Horn, where the train winds round the high brow of a mountain as if it had climbed up to give you a look at the valleys below. The tourist across the Virginias can have delights like these again and again repeated. The Rhine owes no little of its attractiveness to the battlements on its steeps. The New River is not indeed like the Rhine in

Buffalo Gap.

depth or breadth; but it has features of its own. Now it is a broad stream leisurely chattering to the woods that overhang it; anon it is in a narrower bed scolding the rocks as large as houses,

Whale's Head.

that have intruded themselves upon it from the hill-sides, of which they grew weary. But for giant cliffs, Eagle's Nests, Lover's Leaps, and mountain fastnesses in ruins, the New River can compete with any stream of travelled lands, and with this difference in its favor, that no cunning count or baron bold piled up those frowning battlements. Geological forces in an Omnipotent hand, and with unlimited time in which to work, placed these precipitous, castle-like crowns on the wooded hills, and gave them a peculiarity not seen elsewhere, namely, that behind them corn and wine abound; for the Alleghanies are fertile to their summits. As one is whirled along, it is difficult to say which challenges most admiration—the river below, the cliffs above, the graceful lines of the hills, the moving shadows over the green slopes of the mountain sides, or the sublime audacity that built a railroad through such a region. The Chesapeake and Ohio connects with the Virginia Midland at Charlottesville, where passengers from Philadelphia, Baltimore, and

CLIFTON FORGE, C. and O. Railway.

Washington by the latter may change for the Virginia Springs or for points in West Virginia and the West, including Cincinnati and Louisville. A favorite summer route from the East is by water to Richmond or Fortress Monroe and Old Point Comfort, thence west by the Chesapeake and Ohio.

Starting from proud old Alexandria, the Virginia Midland route passes through a section of country not only full of natural attractions, but bristling with points of historical interest dating back to the Revolution and extending down to our late Civil War. Following the southwesterly trend of the Blue Ridge mountains after it leaves Alexandria, the road shows an almost con-tinuous ascent until it reaches the memorable bat-tle-field of Manassas. Here a fine view of the surround-ing country may be had; and from the earth-works, pared down by the hand of time, which mark the out-lines of the entrenched camp built by the Confed-erates, a very wide landscape is seen. At Riverton the Manassas Division of the Midland crosses the She-nandoah Valley Railroad, with its magnificent scenic and metallurgic attractions. Going northward the trav-eller in a few minutes finds himself in Clarke County, and surrounded by the his-toric homes of the gentry of the old days, some of their country seats being on a style that is truly lordly. Washington's office and lodgings at Soldiers' Rest, where Gen. Daniel Morgan, of Revolutionary fame, once lived; Greenway Court, the seat of the ec-

Falling Springs.

centric Lord Fairfax; the old chapel, built in 1796; the homes of Philip Pendleton Cooke, the poet-author of "Florence Vane," and of his scarcely less distinguished brother, John Esten Cooke, the novelist, are in Clarke County. Nor are historic associations with the late war want-ing. many combats and skirmishes having taken place at or near Millwood and Berryville, the county seat. All through this section are various unpretentious summer resorts, where people from the neighboring cities find pleasant homes for the hot months. At Lynchburg close connections are made by the Midland with the Norfolk and Western Railroad for New Orleans, while at Chattanooga divergent lines convey the traveller to Nashville and Memphis, Tenn., Texas, and all points in the South and Southwest.   At Danville the Midland connects with the famous Richmond and Danville system, which under one management extends from the Virginia capital to Augusta and Atlanta, Ga., and embraces over 2000 miles of road.

Among the mineral springs of Virginia and West Virginia are many well-known resorts. The bare enumeration of them all would fill a page or more of this book, and to attempt an account of their curative qualities, or various claims as places of resort, would require the entire space of an octavo volume.   Situated as they generally are, high up among the Alleghany ranges, they enjoy the perfection of mountain atmosphere and an abundance of forest shade. First and foremost are the Greenbrier White Sulphur Springs, in West Virginia, more fully spoken of elsewhere.   The Red Sulphur, also in West Virginia, are twelve miles from Falcott Station, on the C. and O. Railway, and are the only springs of the kind in this country.   The Sweet Chalybeate Springs are nine miles from the railroad, reached by stage from Alleghany,

Lover's Leap, James River, near Lynchburg.

in the height of the mountains.   The "Old" Sweet, as they are called, are ten miles from Alleghany.   These and the Berkeley and Capon Springs are in West Virginia.   Healing Springs, in Bath County, Virginia, are sixteen miles from Covington, over a splendid turnpike.   The Hot Springs are four miles further on in the same locality.   Jordan Alum and the Rockbridge Alum are also on the Chesapeake and Ohio, reached by stage from Gosham and Millboro, respectively.   Rawley Springs are in Rockingham County, twelve miles from Harrisonburg. The Yellow Sulphur are located three miles from Christiansburg, on the Norfolk and Western Railroad, and the Blue Ridge Springs are directly on the line of that road.   The Fauquier White Sulphur are near the terminus of the Warrenton branch of the Virginia Midland.   All these springs are more or less famous and popular, and good accommodations may be found at each.

Coyner's Springs.

THESE White and Black Sulphur Springs are situated on the line of the Norfolk and Western Railroad, in Botetourt County, Va., in the midst of the Blue Ridge Mountains, 46 miles west of Lynchburg, within 5 miles of Roanoke City, and 158 miles from Bristol. The waters of these springs are celebrated for their medicinal qualities, and have been resorted to for years. In cases of difficult, imperfect or painful digestión, enfeebled condition of the nervous system, chronic diseases of the bladder or kidneys, skin diseases, indolent liver, with difficult or vitiated secretions, they will be found to be well adapted. The Black Sulphur Spring is pronounced by physicians a natural emmenagogue, and peculiarly adapted to diseases pertaining to females, and is a specific in most cases. The improvements consist of a large four-story hotel, and cottages ranging on each side of a beautiful lawn handsomely shaded, through which a crystal stream passes. The climate is delightful, and in one of the most healthy situations in the mountains of Virginia. Accommodations for 250 guests. All passenger trains of the Norfolk and Western road stop here, and conveyances from the hotel meet all arrivals. From Washington take the Virginia Midland train to Lynchburg.

Fincastle.

SITUATED six miles from Troutsville, on the Shenandoah Valley Railroad, is the delightful village of Fincastle, which has of late years become a popular summer home for families from the neighboring cities. It is easy of access and the locality is exceedingly healthy, having an elevation of 1200 feet above sea level. Good fishing is reported in the streams and plenty of game near by. By a consolidation of interests a new Union Hotel has taken the place of Hayth's and the Western hotels, and excellent accommodations for a large number are now offered by Mr. William B. Hayth, the proprietor.

Greenbrier White Sulphur Springs.

THESE springs, the most fashionable and popular of all Southern resorts, are frequented every season by thousands of the élite of all sections of the country. The properties of the waters and the surroundings and attractions of the place are fully spoken of in another part of this book, under the head of " Mineral Spring Resorts." Next to the medicinal value of the waters, and the invigorating climate, the company which annually assembles there is most worthy of comment. Statesmen, men of letters, politicians, jurists, belles, and beauties, all gay and brilliant spirits turn to this enchanting spot, and here pleasure takes up her abode. The cottage system, with its pretty homelike surroundings, enables visitors to live in a whirl of gayety or the utmost retiracy, as their wishes may dictate. During the season, besides the nightly balls, there are several grand fancy and masquerades, which add to the amusement of the guests. The accommodations are extensive and comfortable ; besides the cottages, which are a hundred in number, the Grand Hotel is the largest building in the South. The house has been leased for five years by Harrison Phœbus, the popular proprietor of the Hygeia at Old Point Comfort, and it will undoubtedly be kept in the same luxurious style as the latter, achieving a popularity never before attained.

Mountain Lake.

THIS singular and rather attractive place is well known in Virginia as " Salt Pond," but outside of the State there is very little acquaintance with it. The lake, the chief object of interest, is a beautiful and picturesquely situated little body of water, the highest, perhaps, in this part of the country, being over 4000 feet above the level of the sea. It is about half a mile in length by less than a quarter wide, with no visible inlet or outlet. Its depth is something remarkable, reaching, in some places, as authoritatively reported, over 200 feet. The origin of this lake is a mystery, though the traditions of the locality attribute it to the tramping of herds of deer and buffalo frequenting a salt-lick on the spot many years ago, thus causing the earth to " hold water." This explanation, however, would hardly seem to account

Hawk's Nest.

for the great depth of the lake, which has been steadily increasing. Since 1804 this increase has amounted to 25 feet. No drouth ever affects it. The most probable theory is that the presence of this lake is due to some subterranean stream like Lost River. There is some remarkable scenery in the locality, views from the "Crow's Nest" and "Bald Knob," two high points near by, equalling any in the whole range of mountains for extent and beauty. Altogether the place would be one of unusual attractiveness if put in the hands of enterprising owners and provided with better improvements and accommodations. It is reached by stage or private conveyance from Christiansburg on the Norfolk and Western Railroad, or from Eggleston on the New River branch, now open, and twelve miles nearer.

Old Point Comfort.

THERE is scarcely a resort in the country more widely and favorably known than the Hygeia Hotel at "Old Point." Health-seekers from the northeastern and northwestern cities congregate at this half-way place between the tropics and their own colder climate in large numbers during the winter and spring, while many others seek it for the sea air in summer. It is also a favorite stopping-place for thousands who come that way on their return from Florida and the Bermudas. The Hygeia is situated one hundred yards from Fort Monroe, at the confluence of the Chesapeake Bay and Hampton Roads, fifteen miles from Norfolk and Portsmouth. It is reached by daily lines of steamers from Baltimore, Washington, Richmond and Norfolk, and by rail *via* the Midland and Chesapeake and Ohio Railways. The hotel is four stories in height, substantially built and well furnished. It has two Otis elevators, electric bells, with every modern convenience, including hot sea baths. It is in fact a perfect sanitarium. By improvements and additions recently made over 1000 guests can be comfortably entertained at any time. Wide and joyous-looking verandas fronting on the water, having 1500 feet, or about half their extent, encased in glass, during the cooler season, afford retreats where the most delicate may enjoy the sunlight and water-view without exposure. There is music and dancing every evening, and all the pleasures of a fashionable watering-place are to be enjoyed.

Hygeia Hotel, as Enlarged.

The climate of Old Point Comfort is unequalled for salubrity and general healthfulness, malarial fevers being absolutely unknown. The meteorological record for the past ten years shows an average temperature of 74° in summer, 59° in autumn, 44° in winter, and 52° for spring. The whole region roundabout is filled with picturesque scenery, offering delightful drives by day and romantic strolls by night. Boating and fishing are especially attractive, and the surf bathing, which is good from May until November, is unsurpassed on the Atlantic seaboard. For sleeplessness and nervousness, the delicious tonic of the pure ocean air and the lullaby of the waves rolling upon the sandy beach, but a few feet from the bedroom windows, are most healthful soporifics.

A Pennsylvania Forest.—In the Alleghanies.

Pennsylvania Resorts.

VERY American tourist or considerable traveller is more or less familiar with Pennsylvania scenery. In whatever direction one travels over that model railway line which takes its name from the State, but which long since extended itself nearly over the whole country, many glimpses will be obtained of those natural beauties that have been made famous in story and song. One striking feature of Pennsylvania scenery is its endless variety. The entire State is an alternation of mountain ranges, bold cliffs and towering cones, beautiful rivers, charming fertile valleys and rolling landscapes, with here and there a gorge or gap through which a watercourse takes its way to the sea. Those whose eyes are familiar only with the broad prairies of the Mississippi valley or the great plains of the West, will be filled with new emotions upon obtaining a first view of this ever-changing panorama in passing over the Pennsylvania route. The wild and rugged appearance of the mountains, the loftiness of their peaks, and the dense growth of timber covering their sides, is suddenly contrasted with a glimpse of the broad Susquehanna and rich,

The Susquehanna.

highly cultivated farms. The principal rivers of Pennsylvania are almost as well known as the "Father of Waters" in the West. Who has not heard of the Schuylkill, the Delaware, the Alleghany, the Susquehanna, and the beautiful Juniata? And equally world-wide is the fame of such valleys as the Wyoming, the Chester, and the Cumberland. Nor have the beauties of nature been left unimproved by the hand of man. The State teems with a large population, which has covered its surface, particularly along the railways and canals, with great cities and flourishing towns. No pen description can do adequate justice to the surface appearance of the Keystone State, a section of the Union which all Americans will find it profitable and enjoyable to behold before searching for pleasure in the Old World. To the New Englander the State of Pennsylvania is as much of a surprise as it can be to the farmer from Missouri or Kansas. He finds Philadelphia considerably larger than Boston; he sees colleges, churches, and schools in every direction, which equal those he left at home; he crosses rivers much wider and longer than the Connecticut or Penobscot; he discovers that the Green Mountains are a row of hills by the side of the Alleghanies; and he looks with

wonder, if not astonishment, upon the coal and iron mining operations.   Railway trains run
into the sides of lofty mountains; they pass under ranges of mountains; they scale their very
tops, and run up and down the sides of the steepest with no apparent friction.   The feats of
railway engineering which have been accomplished in Pennsylvania are second only to those
we read of in the Andes of South America, the Alps of Switzerland, or the Rockies of
Colorado.

Mill Creek, Penna. Railroad.

But this State is not all mountains, mines, and railways.   Portions of it, as in the vicinity
of Harrisburg, York, and Philadelphia, seem as if the garden spot of the Union.   The farms
are immensely productive; the barns are large and bursting; the dwellings handsome, modern
and commodious.   If Pennsylvania were in Europe it would rank as a first-class kingdom, the
envy of its neighbors.   At various points in the State adjacent to the Pennsylvania Railroad,
are many natural curiosities of considerable note.   In the neighborhood of Lewistown, a
beautifully located spot, are several curious caves.   Alexander's, in Kishicoquillas Valley,

abounds in fine stalagmites and is a natural ice-house, preserving in the midst of summer the ice formed in winter. Hanewall's Cave near McVeytown is of vast dimensions, and contains calcareous concretions. Near Tyrone is Sinking Spring Valley and the creek from which it takes its name. This creek emerges from the Arch Spring, and then proceeds to lose itself again and again as it flows onward. Some of the pits through which it is visible are several hundreds of feet in depth. Many of these openings are seen along the sunken stream, which at length appears upon the surface for a short distance. It then enters a large cave, through which it flows in a channel about twenty feet wide for a distance of more than three hundred yards, when the cave widens, the creek turns, and is plunged into a cavern where the waters are whirled and churned with terrific force. Sticks and long pieces of timber are immediately carried out of sight, but where they go has never been ascertained—no outlet for the waters having been discovered. This curiosity is much visited by parties from the neighboring resort, Altoona. Another peculiar formation is "Jack's Narrows" near Mt. Union, made by the river forcing itself through Jack's Mountain. This gorge is wild and rugged in its appearance, the sides being almost destitute

Jack's Narrows, Penna. Railroad.

of vegetation, exposing immense masses of gray and sombre rock. The mountain receives its name from a weird, mysterious hunter and Indian slayer, who made his haunts in the valley previous to the Revolutionary War. The Narrows were called in early Colonial records, "Jack Anderson's Narrows," from the fact that in them an Indian trader named John Anderson and his two servants were murdered by the savages.

Some of the finest scenery in Pennsylvania is in the Alleghany Valley, partaking of the peculiarities of the "beautiful river" of the early French explorers. This stream is remarkable in many respects. By means of French Creek and Le Bœuf Lake, and Conewango Creek and Chautauqua Lake on the northwest it almost touches Lake Erie; on the northeast it stretches out its long arms towards the Genesee River in New York and the north branch of the Susquehanna, while on the south it pours its waters through the Ohio and Mississippi into the Gulf of Mexico. For the greater part of its course it flows, not through a broad valley

Alleghany River at Freeport.

like most rivers, but in a great ravine, from one hundred to four hundred feet below the level of the adjacent country. The scenery is in some places of the wild and rugged sort, but more generally is picturesque and beautiful. The hills though steep are clothed with a dense forest, presenting the appearance of vast verdant walls washed at their base by the limpid waters. There are no rocks, strictly speaking, in the channel. But the most famous portion perhaps of the Keystone State, is the valley of the Schuylkill, through which runs the Philada. and Reading Railroad. The praises of this beautiful river and its flowery banks were sung by the poet Moore, who many years ago occupied a lowly cottage near what is now Fairmount Park. The valley is also historically famous for having been the scene of some of the darkest episodes of the Revolution. Here, amid the snows of December, Washington with his little army of frozen, barefooted patriots was threatened with attack by the British commander, Howe, and his force of 14,000 redcoats. Skirmishers were frequently out, but on the 16th of that month the invaders began to draw in their lines, evidently satisfied that their opponents were too strong. Finally both armies went into winter quarters, Howe at Philadelphia and Washington at Valley Forge down the river. Among the Generals in our army there was quite a discussion as to whether quarters should be taken up at Reading, York, or Carlisle, and the result is thus stated by one of the poets of that or a later period :

> " But Washington decided
> When all had spoken round,
> That Valley Forge, in Chester,
> Should be our winter ground."

Near here is Phœnixville, a prettily situated town, which has the honor of having produced the iron of which the dome of the capitol at Washington was made. Norristown, Pottsville and Reading, three of the important inland towns of the State, are also situated in this Valley. The original line of the Philadelphia and Reading Railroad extended only between the two cities from which it is named—a distance of 58 miles—but the road now comprises, including branches and leased lines, over 1500 miles of track. It runs through valleys and up mountains, in all directions, and embraces some of the finest scenery to be found anywhere. It also forms in connection with the Central Railroad of New Jersey, the popular Bound Brook Route between Philadelphia and New York. Lovers of the picturesque find enough to interest them in the vicinity of Reading, and further on near Quakake Junction the Railroad climbs inclined planes up the sides of the highest mountain, affording views of landscape unsurpassed in extent and beauty. Here we see the typical American forest in all its wildness. Beyond, across the narrow valleys, is Catawissa creek, rolling and lashing along its rocky channel.

A Glimpse of the Schuylkill.

The summer resorts of Pennsylvania are not for the most part to be classed as fashionable resorts such as are sought by those votaries of the giddy goddess who go abroad in summer to dress and dance and to keep up the dizzy whirl in which they have lived all winter at home in the city. But for those who desire rest and change, with pure air and the enjoyment of nature, they are peculiarly adapted.

Altoona.

Few places in Pennsylvania present more genuine attractions, considering both health and pleasure, than this. Situated at the head of Logan Valley, on the main line of the Pennsylvania Railroad and on the western slope of the Allegheny mountains, it possesses many special advantages in the variety and extent of its surrounding attractions and the number of inter-

esting objective points for drives or short trips by rail.   Altoona is the Summit City of Penn.
sylvania, being 1200 feet above the level of the sea and in an atmosphere of unusual purity,
under the influence of which asthmatic sufferers and the victims of hay fever, in many cases,
find immediate and complete relief.   The scenery of the locality is of the most varied descrip.
tion and presents, within a radius of a few miles, a gradual transition from the graceful and
picturesque to the rugged and sublime.   A short distance west is the famous ‘‘ Horseshoe
Curve.’’   The valley here separates into two chasms, but by a grand curve, the sides of which

Logan House, Altoona.

are for some distance parallel with each other, the road crosses both ravines on a high em.
bankment, cuts away the point of the mountain dividing them, and sweeps around and up
the stupendous western wall.   Looking eastward from the curve, the view is peculiarly im.
pressive, while at Allegrippus, where the majesty of the mountains seems to culminate, the vast
hills in successive ranges roll away in billowy swells to the far horizon, the prospect being only
bounded by the power of vision.   Twice each day during the summer open ‘‘ observation
cars’’ are attached to the day express trains, and make the round trip between Cresson and
Altoona, enabling passengers to see with ease and pleasure the unsurpassed scenery of the
Alleghanies.

Opportunity is afforded for another pleasing diversion by the vicinity on the north of the Wopsononoc mountain, easily accessible to carriages, from whose summit is spread before the eye a panoramic view which is, in the opinion of experienced travellers, unsurpassed upon either continent in all those features which delight and inspire. It comprises the entire valley of the "Blue. Juniata," a picture of highly-cultivated farms and smiling peace and plenty, bounded by swelling ranges of hills, which gradually fade away in the azure of the distant horizon. The celebrated "Sinking Spring Valley," with its subterranean streams and immense caverns, lies to the eastward, while on the southeast is the Bell's Gap Narrow-Gauge Railroad, excursions by which, to the summits of the mountain, are among the most satisfactory and popular diversions of life at Altoona. The views in this locality are less extended and open. The valleys become huge ravines, from which the hills rise on either side almost precipitously. The grade of the road rises one hundred and fifty feet to the mile, and as the diminutive trains creep up and along the sides of the vast amphitheatre of living green, the scene is such as to defy the power of pen description. To the facilities of the Logan House for supplying the "creature comforts" no elaborate allusion is necessary. The building itself, surrounded by broad piazzas, is elegant in all its appointments and provided with all conveniences, including electric bells. The elevated site, charming surroundings, delightful air, and convenience of access combine to render it one of the most desirable resorts in the State. The large and beautifully shaded lawn affords a fine field for croquet and other outdoor sport, while within ten-pin alleys, billiard tables, etc., provide ample facilities for recreation. All the mountain streams in the vicinity abound in trout, rendering the locality a paradise for the angler. Altoona is but eight hours' ride from Philadelphia and Baltimore, nine from Washington, ten from New York, and three from Pittsburgh. Passengers from these points are assured of transportation facilities of the most perfect character, via the Pennsylvania and Northern Central Railroads. The traveller by this line who regales himself at the Logan House, on his journey, will see conspicuously painted upon the wall of the great dining-room, a picture representing, in all the gorgeousness of savage dress, Logan, the famous Mingo chief, whose name is associated with the earlier history of the State.

Bedford Springs.

THESE springs, situated in Bedford County, one mile from the town of Bedford, enjoy a high reputation for the health-restoring qualities of their waters and the air of the locality. The waters are recommended for a wide range of diseases, including those of the liver, the kidneys, and the skin, and for some of these ailments are pronounced absolute specifics. A distinguished physician, writing to the "Medical Examiner," says: "I have, myself, twice gone to Bedford so prostrated as scarcely to endure the fatigue of the journey, and wholly disqualified for all exertion, and have in both instances returned, at the end of a fortnight or three weeks, restored to my wonted power of labor, and have witnessed similar results in the cases of friends and patients." The springs were discovered in 1804, and the following year were frequented by persons afflicted with diseases, who encamped in the valley to be near the newly-discovered fountain of health. Not long afterwards accommodations were provided for visitors, and for threescore years they have regularly drawn a large number of health and pleasure seekers. The natural beauty of the valley where the springs burst forth is great, and it seems to have been formed by nature as a retreat for wearied and suffering humanity. High hills surround it, ascended by terraced walks, and from their summits pleasing vistas open. From the elevated position of these springs, among the ranges of the Alleghany Mountains, and the dense forest growth surrounding them, the atmosphere is always deliciously

cool; and doubtless much of the benefit derived by visitors is owing to the fact that no suffer_
ing is experienced from a midsummer sun, and that refreshing sleep can always be enjoyed.
Bedford is an old town, and has an interesting history.   It was the site of an important fort

Bedford Springs.

in colonial times, and some of the most illustrious names in American annals are associated
with events occurring here towards the close of the eighteenth and in the early years of the
nineteenth centuries.   The adjacent country is picturesque—fertile valleys and rugged moun_
tains, holding rich deposits of iron-ore, abounding in all directions.

Cresson Springs.

AMONG the Pennsylvania Resorts which have attained great popularity, Cresson stands second to none. It first became famous for the curative properties of its mineral springs and the exceeding beauty of its surroundings, combining all the attractions of a quiet mountain resort with the advantages of superior medicinal waters, and adding to the enjoyment of both the excellent accommodations and conveniences of the best city hotels. Cresson is located almost on the summit of the Alleghanies, 2300 feet above the level of the sea, in the midst of the most delightful scenery, and while thus affording irresistible attractions for the heat-oppressed and care-worn seeking a quiet retreat where rest and recuperation may be had without the sacrifice of personal comfort, it is easy of access from all the great cities, being immediately on the main line of the Pennsylvania Railroad within a few hours' ride from New York, Philadelphia, Baltimore, and Washington, in the East, and Pittsburgh, Cincinnati, Indianapolis, and Louisville, in the West. Round trip excursion tickets from all points to and from Cresson are on sale at all ticket offices during the season. The mineral springs at Cresson flow from the mountain in the vicinity of the hotel, and it is attested by eminent medical authority that "there are no more valuable medicinal waters in the Union than those of Cresson Springs." The water of one has aperient action, while another possesses decided tonic properties. The alum water is one of the most valuable agents known for loss of tone and vigor to the skin, in general debility, and in all congestive conditions of the skin. In the centre of the hall of the hotel, set up in capacious coolers for the free use of guests, are all the mineral waters of the place, with those of Saratoga, Bedford, and Minnequa.

The new MOUNTAIN HOUSE, erected during the fall and winter of 1880-81, on the site of the old hotel, is a very striking structure in the Queen Anne style of architecture, into which is blended the Oriental. It is located on the crest of a hill in the midst of a delightful grove. The main front is 300 feet long, with an elevation of about 100 feet, embracing four stories and basement, with wings extending from each end to a depth of 220 feet. The whole building is surrounded by a covered piazza 16 feet wide, forming a promenade 1200 feet in length. The new hotel, although having accommodations for seven hundred guests, was found inadequate to accommodate one-half of those desirous of patronizing it last season, and during the past winter it has been enlarged by an extension of the west wing to the dimensions contemplated in the original plan—the addition containing eighty sleeping-rooms. At the eastern end another wing has been added, containing ten-pin alley and children's dining-room, the room used for the last-named purpose last season having been incorporated in the main dining-room, increasing its seating capacity to fully 800. The hotel itself will accommodate about 900 guests, in addition to which there are 25 cottages in the surrounding grove, providing special accommodations for those seeking seclusion and the perfect quiet of home, while dwelling near enough to the concert and the dance to participate at pleasure. The cottages form a portion of the Mountain House property, and are managed by and under the immediate charge of the hotel officers and servants, meals being served either in the main dining-room of the hotel or at the cottages, as specially arranged. Board walks extend from the main building to all the houses, and the ways are brilliantly lighted at night. As the railroad station is at the foot of the lawn, within a stone's throw of the main entrance, no fatiguing stage or wagon ride at the end of a long journey is necessary in going to or from Cresson.

In contrast with the mountain fastnesses all around Cresson the beautiful and extensive grounds about the hotel have a peculiar charm, there being about 400 acres of land in lawns, gardens and groves. The surroundings of the hotel are attractive, and pleasant drives lead away through the almost unbroken forests, where the laurel, the hemlock and the pine afford

MOUNTAIN HOUSE, CRESSON.

(34)

a delightful shade and fill the air with the ceaseless rustle of their branches. A mile or so from the house is the Old Portage Road, with its ten inclined planes, by which the Pennsylvania Road originally crossed the mountains. It was once one of the wonders of this continent, but is now abandoned, and is visited only by the curious and the student of our system of internal improvements. Comfortably seated behind one of the fine teams always to be depended on at the livery, connected with the office of the hotel by telephone, the drive over the Old Portage Road is one of the finest through wild-wood scenery the visitor can take. Ferns and wild flowers grow on all sides, beautiful vistas through the trees greet you at every turn, and the smell of the green spruce and pine foliage, deepened by the dew and borne on the cool air, is delicious.

Delaware Water Gap.

THIS peculiarly beautiful and picturesque resort is known to tourists far and wide throughout our land. The name "Water Gap" is given to that point in the course of the Delaware river where it forces its way through the Kittatinny or Blue Ridge Mountains. Mounts Minsi and

Delaware Water Gap.

Thompson House, Kane.

(36)

Tammany form the walls of the Gap, their almost precipitous sides rising against the horizon to a height of a thousand feet, approaching each other closely as if in determination to bar the river's course. Indeed, it is believed they did so at one time in the thousands of years agone. The Indians gave to the valley north of the Blue Ridge and above the Gap, the name of "Minnisink," or "Whence the Waters are Gone." "Here," says a writer, "a vast lake once probably extended, and whether the great body of water wore its way through the mountain by a fall like Niagara, or burst through a gorge, or whether the mountains uprose in convulsion upon its margin, it is certain that the Minnisink country bears the mark of aqueous action in its diluvial soil, and in its rounded hills, built of pebbles and boulders."

The attractions of the Delaware river, which, above Trenton, is one of the most picturesquely beautiful streams in the United States, culminate at the Water Gap, and form a location equalled by few in the country in its adaptation to the purpose of health and pleasure. An organization of gentlemen from New York and Philadelphia, yclept the "Minsi Pioneers," through a long course of systematic and well-directed labor, have opened a great number of paths and rambles upon the mountain side, and have thus added a feature to the other attractions of the Gap which is of inestimable value. These rambles are practically exhaustless in fine views and situations, and all along the route are scattered seats and rustic summer-houses for their pleasant contemplation. The summit of Mount Minsi is easily accessible to carriages, and from its narrow crest, scarcely more than fifty feet wide, a panoramic view may be obtained of vast extent and varied and unexcelled beauty. Prominent among the special points of interest, and which afford objective points for a pleasant ramble, are Eureka Glen, famed as the favorite of George W. Childs, and rendered accessible through his liberal expenditures by a succession of rustic bridges and stairways, and Moss Cataract, Diana's Bath, and Caldeno Falls, located on Caldeno Creek, a little stream which takes its rise in the Hunter's Spring, a cool and sequestered spot far up in Minsi Mountain, though quite easily reached by a path.

In the social life at the Gap there is none of the gayety and excitement which characterize our sea-side resorts, and there are few allurements for the votaries of fashion. All that is best, however, in representative American people is fairly represented. The daily life and occupation are conspicuous in the absence of all conventional restraints, and are characterized by as much freedom as life in a country farm-house, while, at the same time, the hotel accommodations provide all comforts and conveniences. Delaware Water Gap is one hundred and eight miles distant from Philadelphia, and ninety-two miles from New York. It is reached from the former place by the Belvidere division of the Pennsylvania Railroad, and from the latter *via* the Morris and Essex Railroad.

Kane.

LOCATED in the wildest portion of Pennsylvania, on the highest summit of the Alleghanies reached by the Philadelphia and Erie Railroad, Kane possesses superior attractions for sportsmen and for those who are inclined towards life in the forest, with all the comforts of a good hotel at the same time. Its elevation and dense hemlock and pine forest surroundings give it an atmosphere of peculiar rarity and healthfulness, very beneficial in cases of asthma, hay fever, and other diseases of the respiratory organs. "The country adjacent to this station is celebrated for the production of milk, butter and cheese—the manufacture of the latter article being an important industry, prosecuted on an extensive scale. The markets are abundant—game, mountain trout, and the luscious fruits of the forest being obtainable in any quantities when in season. Sulphur and iron springs burst forth near the hotel, and throughout all the region limpid streams and pools abound, filled with the speckled trout so attractive to fishermen, and in some of which the breeding and rearing of these beauties is scientifically

Renovo Hotel.

(38)

carried on. The forests, almost interminable in extent, are intersected with good dry roads, carpeted by the cast foliage of hemlocks and pines, and arched by their perennial verdure, where drives and walks afford unalloyed enjoyment. Deer are frequently seen browsing on the herbage or bounding through the woods; rabbits scamper along the roads; pheasants awaken the echoes with their drumming, and silent woodcock whirl away from approaching humanity to seek more secluded retreats. Sportsmen can always procure safe, experienced guides to pilot them where guns and rods can be brought into active play, and the lover of nature can find many places to pause at, and scenes to remember.'' The "Thompson House," at this station, is a large and very superior hotel, elaborately furnished, combining all the comforts of the finest summer resorts. Kane is reached from Washington and Baltimore via the Northern Central Railroad, connecting at Williamsport with the Philadelphia and Erie. From Philadelphia and New York take Pennsylvania Railroad.

Renovo.

ANOTHER of the seductive spots in the Alleghany forest, on the line of the Philadelphia and Erie Railroad, is Renovo, which has lately become a popular resort, ranking with Kane and Altoona. It is delightfully situated in a little oval valley, formed by a separation of mountain ranges rising around it to a height of more than a thousand feet, through which the West Branch of the Susquehanna river glides in a placid and pellucid current. It is the location of the railroad workshops, and the industry centred here by these improvements is the life stimulus of the place, causing it to grow, in a period of about ten years, from an isolated farm into a town of more than two thousand inhabitants. The scenery in the vicinity is charmingly picturesque—in some localities rising to sublimity and grandeur. Renovo may be said to lie almost in the heart of the great pine forests of Pennsylvania, and the depth of those mysterious woods where the sportsman will find ample uses both for the gun and the rod, can readily be reached from it. The hotel at Renovo, erected and owned by the railroad company, is large and comfortable, affording accommodations unsurpassed in excellence. Directions for reaching Renovo are the same as for Kane.

Mauch Chunk.

THIS is an extensively advertised resort in the coal regions of Pennsylvania. Its chief attraction is what is termed a "Switch-back" railroad up a small mountain, the cars being drawn by a stationary engine. The scenery round about is more or less interesting, according to taste. The Lehigh valley, in the vicinity, is generally regarded one of the picturesque portions of the State. There are two moderately comfortable country hotels in the place, both of which take summer boarders.

Our Natural Wonders.

" Thus Nature works as if to mock at Art,
And in defiance of her rival powers;
By these fortuitous and random strokes,
Performing such inimitable feats,
As she with all her rules can never reach."

ATURE in planning and fitting the theatres of the world was never more lavish of fine and startling effects, and yet never more harmonious, than when her hand placed the scenery amidst which the nations of the New World are playing their part in the great drama. In this land of wonders are amphitheatres whose seats rise in tiers of circling ranges of mountains till lost in the clouds, with sweeping arenas on which Time may place the great tragedies of the hereafter. Here are all the accessories at their best and pleasure-spots innumerable; here waterfalls where seas have thundered over mighty walls, and others where the stream leaping from a giddy height floats down a rainbow-tinted Bridal Veil; here caves which might hide all the bandit bands from Ali Baba's down, and with glittering splendor of gorgeous chambers and sculptured grottoes surpass even the fabled homes of the sea-kings; immortal statues adorn the western platforms, and grand countenances cut from granite rocks look down upon the eastern plains. Romance may revel in wonders where Nature has draped the rocky walls and where sunlight filters down through green-bowered banks into glens and chasms filled with sparkling cascades, purling brooks and mirror pools; while the supernatural finds an appropriate ground where inky-black water boils and bubbles, where streams of hot water pour over barren rocks, where steam and gas and sulphurous fumes spit up in jets from the cinder-covered crust, where howling, tearing columns of steam and scalding liquids burst from the bowels of the earth and fall booming and rumbling back to the flame-circled throne of Satan.

Where flow the mighty rivers beyond the Mississippi are reared the citadels of Nature. Here over thousands of square miles are mountains and architectural piles, stately and grand. Castles and abbeys, columns and obelisks, bastions and donjon towers, minarets and peaks, turrets and spires lie strewn along the course of an eddying, muddy, swift-rolling current, pent in between walls a third of a mile high, over which at intervals a newer stream drops, to fall affrighted and trembling on the rocks below. Everywhere perpendicular cliffs, the surface worn and eroded, polished curves and cunningly cut figures in varying shades of yellow and red. One region bare, rugged and awful; another interspersed with parks and gardens which shame the attempts of man. In many a picturesque portion the colors of the soil and sandstone blend as if laid on by a great master; mullioned peaks hundreds of feet above show a dreamy glimpse of the far sky; thousands of twittering birds find happy homes in the numerous percolations in the tremendous cliffs; while as an ideal carpet or a gorgeous rug over wide fields are spread bowers and beds of flowers, whose beauty, grace and color art can never capture or words picture. Clumps of green, oak bushes, thickets of wild roses, tiny tufts of fern, the sweet clematis and the bright woodbine climbing down the very brows of the beetling precipice and nodding a gay greeting to the blue violets below; the red lily and the yellow columbine; the white spiræa and the mertensia with its pink and blue bells; the kinnikinnick and the splendid, many-budded penstemon; the glorious variegated gilia and the dear, familiar daisy. All these and a hundred others, the final loving touches to the marvellous works which, hoary from the hand of Time, are yet all but new to the eye of man. As a worthy addition to these wonders of our home are those ancient, sad and sombre groves,

(40)

stupendous pine cathedrals that never shadowed the shrine of a Druid or murmured amongst their sighing boughs the dread oracle of a deity.

The Yellowstone National Park.

THE title of "Wonder Land," which has been bestowed upon that immense national reservation known as the Yellowstone Park, is aptly applied. No such aggregation of all that is wonderful in nature exists anywhere else on this continent, if in the world. The American tourist who goes abroad in search of curious things and natural delights before he has visited this marvellous region commits an unmistakable error. The Yellowstone National Park is situated along the highest part of that great culminating area of North America which has been fitly termed "the crown of the continent." It lies, for the most part, in the northeast corner of Wyoming Territory, between the 44th and 45th parallels of latitude, and reaches from the 110th meridian to a short distance beyond the 111th, extending on the west and north for a few miles into the adjacent Territories of Montana and Idaho. It is a region sixty-five miles long by fifty-five miles wide, and covers an area of about three thousand five hundred and seventy-five square miles; or, to give a still clearer idea of its extent, it may be said to be large enough to more than contain the States of Delaware and Rhode Island. This tract of land has, by an Act of Congress, been "reserved and withdrawn from settlement, occupancy or sale under the laws of the United States, and dedicated and set apart as a public park or pleasure-ground for the benefit and enjoyment of the people" forever. The mountain ranges which hem in the valleys, rise to a height of ten and twelve thousand feet, and are covered with perpetual snow. The average elevation of the country included in the park is about six thousand feet, which is as high as the summit of Mount Washington in the White Mountains. A fine general view of the Park is obtained from the summit of Mt. Washington, a central accessible position.

It would not be possible to adequately portray in these pages all the wonders embraced within the confines of the Yellowstone National Park. Its most striking features are its geysers, hot springs, waterfalls and cañons. In number (ten thousand, it is said,) and magnitude of its

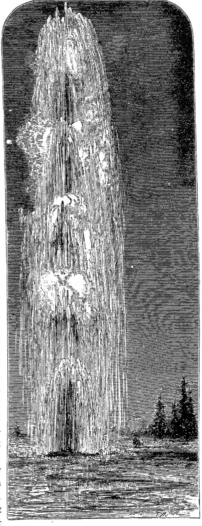

The Giantess.

hot springs and geysers it surpasses all the rest of the world. The geysers are the grandest in the universe — grandest in the frequency of their eruptions, in the quantity of water they spout, and the height to which it is thrown, and also in the beauty of their delicately ornamented and often brilliantly-colored chimneys and basins, built up and adorned by the minerals deposited from their hot silicious waters. They spout columns of boiling water of sizes varying with the dimensions of their orifices—from a few inches to twenty feet in diameter—and to heights ranging from two hundred and fifty to two hundred and seventy-five feet, the eruptions being accompanied by a constant succession of miniature earthquakes, by a terrible noise like almost continuous underground thunder, and by the evolution of immense masses of steam, which roll up wreath after wreath hundreds of feet above the water. These indescribably magnificent displays occur with some geysers at fixed periods, as in the case of the Old Faithful, which spouts from an orifice seven feet long by two feet wide every sixty-five minutes, its eruptions lasting from four to six minutes. It is the only geyser in the world which spouts so fre-

Lower Falls, Yellowstone.

quently and with such unfailing regularity, whence its name. The most prominent geysers are the Giant and Giantess, Old Faithful, the Beehive, the Old Castle, the Grand, Riverside, Comet, and Fantail.

Next to the geysers, as matters of attraction, may be reckoned the Falls and Grand Cañon of the Yellowstone River. This river is a tributary of the Missouri. Its nominal source is the Yellowstone Lake, though its real beginning is the head of the Upper Yellowstone, about twenty-five miles further up in the mountains. The vicinity of the source of the Yellowstone marks a point from which pour down to the Gulf of Mexico on the southeast, the Gulf of California on the southwest, and the Pacific slope on the northwest, the mightiest rivers of both coasts of the continent. About fifteen miles below the lake are the

The Lower Canyon.

Upper Falls (one hundred and forty feet abruptly), while the Lower Falls (three hundred and ninety-seven feet high, which is two hundred and twenty-six feet higher than Niagara), are something over a quarter of a mile farther down, and at the head of the Grand Cañon, whose brilliantly-colored portion begins near the picturesque Crystal Cascades of Cascade Creek,

The Grand Canyon.

which, about midway between the two falls, leaps over the west wall of the cañon in beautiful cascades, one hundred and twenty-nine feet high. These cataracts and cascades would justly rank among the great wonders of the world if they poured into the most commonplace of gorges, but they plunge into an abyss more than twenty miles long, with walls from one thousand to three thousand feet high, and perhaps more, for the lower part of the Grand Cañon has never been explored. This stupendous channel through the Elephant's Back mountains is all the way cut through soft volcanic rock, which has eroded into innumerable quaint forms. But wonderful as these falls are for their height, and the curious forms into which they have weathered, they are vastly more wonderful because, from the water's edge to the top, nature has dyed them with an endless variety of the most vivid colors. Surely nothing like it was ever seen out of fairy land.   Cliffs half a mile and more in height stretch farther than the eye can reach, a mass of yellows, from gold to pale straw ; of reds, from deep carmine to softest pink, everywhere intermingled with coal-black and snow-white, and cream, and buff, and brown, and gray, and olive, and russet, the pure blues, greens, and blue-purples being supplied by the clear sky, by the patches of vegetation growing in places

Upper Falls of the Yellowstone.

on the gentler slopes, by the evergreen trees clinging here and there along the walls or . crowning the platter-topped towers, and by the broad pools in which the river lazily whirls and rests between the succession of cascades down which it dashes a mass of snow-spray, so shrunken to the naked eye by the enormous depth that it often seems a mere silver thread strung with emeralds.

It is beyond the power of language to portray the marvellous grandeur and beauty of the Grand Cañon. It has no parallel in the world. Through the eye alone can any just idea be gained of its strange, awful, fascinating, unearthly blending of the majestic and the beautiful; and, even in its visible presence, the mind fails to comprehend the weird and unfamiliar, almost incredible scenes it reveals. At the foot of the Grand Cañon Tower Creek empties into the Yellowstone. This wonderfully beautiful stream has its rise in the high divide between the valleys of the Missouri and the Yellowstone, and flows for about ten miles through a cañon so deep and gloomy that it has earned the appellation of " Devil's Den." About two hundred yards above its entrance into the Yellowstone the stream dashes out from among jagged pinnacles and massive towers of almost black amygdaloidal lava, and pours over an abrupt descent of one hundred and fifty-six feet, forming one of the most unique and beautiful cataracts to be found in any country. These falls, which are about two hundred and sixty feet above the Yellowstone at the mouth of Tower Creek, are known as Tower Falls, taking their name, as does also the creek, from the columns of volcanic breccia surrounding them. Some of these columns resemble towers, others the spires of churches,

Yellowstone Lake.

and others shoot up as lithe and slender as the minarets of a mosque. So sharply cut are some of these pinnacles and towers that one can scarcely believe that they have not been chiselled by man.

It must not be thought that the waters in the Yellowstone National Park are always dashing and splashy, and rushing and roaring like the water that comes down at Lodore. The country embraces lakes whose shores run off into hundreds of miles, and whose surfaces are as clear, placid, and beautiful as any on earth. Yellowstone Lake, secluded amid the loftiest peaks of the Rocky Mountains, possesses strange peculiarities of form and beauty, and is one of the most attractive natural objects in the world. This beautiful sheet of water is large enough to float all the navies of the world, its superficial area being about three hundred square miles, and its greatest depth three hundred feet. Its elevation above the sea is seven thousand four hundred and twenty-seven feet, which is but little lower than the high lakes of Colorado and Lake Titicaca in South America. It receives no tributaries of any considerable size, its clear, cold water coming from the snows that fall on the lofty mountain ranges that hem it in on every side. In the early part of the day, when the air is still and the bright sunshine falls on its surface, its bright green color, shading to a delicate ultramarine, commands the admiration of every beholder. Later in the day when the mountain winds come down from their icy heights, it puts on an aspect more in accordance with the fierce wilderness around it. It contains several beautiful islands, and is of so irregular form as to give an uncommon beauty alike to its bold, bluff shores and its stretches of sandy, pebbly beaches. Its waters swarm with salmon-trout, and are the summer home of countless swans, pelicans, geese, brants, gulls, snipe, ducks, cranes, herons, and other waterfowl, while its shores, sometimes grassy, but generally clothed with dense forests of pine, spruce, and fir, furnish coverts and feeding-ground for elk, antelope, black and white-tailed deer, bears, and mountain sheep. Scattered along the shores of the lake, and on the mountain slopes which overlook it are many clusters of hot springs, solfataras, fumaroles, and small geysers. At one point a hot spring, boiling up in the edge of the lake, has deposited the mineral carried in solution by its waters, and built up a rocky rim about itself, so that wading out into the lake you can climb on the rim of the spring, and standing there can catch trout out of the cold water of the lake, and without moving from your tracks, can turn round and cook them in the spring.

Twelve miles down the Yellowstone the "Devil's Den" forms an entrance, like a hallway, into a rock wall, from which flows boiling, sulphurous water. Clouds of sulphurous, suffocating steam constantly puff from this opening. Here too, "Mud Volcano" roars like a tempest, and flings hot mud in all directions during an eruption. Near the "Devil's Den" is Brimstone Mountain, where pure sulphur may be shovelled by the cart-load.

First Boat on Yellowstone Lake.

The Hot Springs of Gardiner's River are to be reckoned among the great wonders of this marvellous region. They are among the first things to claim the tourist's especial attention on entering the Park from the northwest from Bozeman, Fort Ellis, and the Bottler Brothers' ranch. Below Tower Falls and the mouth of the Grand Cañon, at the lower end of what is known as the Third Cañon, Gardiner's River, a mountain torrent twenty yards wide, cuts through a deep and gloomy gorge and enters the Yellowstone. At this point the Yellowstone shrinks to half its usual size, losing itself among huge granite boulders, which choke up the stream and create alternate pools and rapids swarming with trout. Worn into fantastic forms by the washing water these immense rock masses give an aspect of peculiar wildness to the scenery. But the crowning wonder of this region is the group of hot springs on the slope of a mountain four miles up the valley of Gardiner's River. The level or terrace upon which the principal active springs are located is about midway up the sides of the mountain. "G. H. B.," the famous newspaper correspondent, who spent the latter part of April and first days of May of this year in the Yellowstone Park, writes as follows of the Hot Springs and other grand and grotesquely beautiful freaks of Nature in this region: "By an ascending pathway we reached a plateau of dead-level tableland, from which is gained a full view of the sixty-seven springs in the valley. White terraces frosted with the salt crystals gleam in the sunshine. The terraces are fourteen in number, rising from the river. Here and there columns of steam and jets of boiling water are flung high in mid-air from their caldrons below. Singularly, the hottest spring is found upon the highest terrace. The gorge in which they are located is 1000 feet above the level of the Gardiner River. These springs, bubbling and boiling in their basins, have washed deposits of lime from terrace to terrace, forming great reservoirs rimmed with delicate lime lacework, and hung with strands of seeming ivory beads; crystal formations in brilliant coloring, luminous tints and shades, terraces and pools frosted like lacework, keep the eye dancing and the imagination at fever heat. As if placed there by art, a central spring adorns each terrace. These are surrounded by a large basin, over the rim of which the water flows down the declivity, forming hundreds of reservoirs, their margins fringed with exquisite tracery, like lace and bead-work. To these add glittering stalagmites, stalactites, and grottoes, painted with the colors of the rainbow, brilliant and fresh, and you are in the presence of a scene such as the world nowhere else holds.

Hot Springs, Gardiner's River.

The drawbacks to visiting this interesting region are twofold: the tediousness and great expense of the trip, and the lack of accommodations when there. It is a long journey from any part of the East to Yellowstone Park, and rendered more difficult, tiresome and expensive by reason of the long stage-ride beyond the railway terminus. By way of the Union Pacific route tourists are obliged to change at Ogden to the Utah and Northern Railroad, which takes them to Beaver Cañon. From this point there is two or three days of staging over a distance of 110 miles to the Firehole Basin. Stage fare for the round trip is $25. By the North Pacific route the railway stops at Livingston, whence there is also two or three days' staging to the northern boundary of the Park. When all this is overcome there is little chance of living comfortably while seeing the wonders of the locality, except for parties who combine in sufficient strength to provide their own commissary and camping facilities, which involves great expense. The scheme of Uncle Rufus Hatch and his company of speculators to get possession of the Park with enormous privileges was partially frustrated by Congress. The company, however, did obtain from the Secretary of the Interior a lease of ten acres, in selected tracts, upon which hotels will be erected some time this year. This, with the further extension of the North Pacific road towards the Park may make the tour one of somewhat less difficulty after this season.

Niagara Falls.

PROBABLY a larger proportion of people have seen Niagara than any other "wonder" of this continent. It has been admired, and written about, and wondered at, since the American colonies were first settled until to say anything new or fresh concerning it has become a task of exceeding difficulty. In a lecture recently delivered in New York, Rev. Robert Collyer stated that he had in his possession a book written by an Englishman, describing his visit to the falls eighty years ago. Such was the world-wide fame of the wonderful cataract that no foreign tourist, even then, considered that he had seen one-half the wonders of this wonderful country without a visit to the falls. The trip up the Hudson took four days, and the writer of the book had among his fellow-tourists an aristocratic federalist—for there were aristocrats in those days—and a clergyman, who was very drunk. It took from the 2d of July till the 23d of August to reach the falls, and the end of the trip was made by following an Indian trail. The grandeur of the falls compensated for all the pains, and was the only thing the irascible Englishman did not grumble at during the long journey. Such was Niagara at the beginning of the century.

Niagara Falls.

This great cataract cannot be described. Its dimensions may be given—its height, and breadth, and volume told—but still much is lacking. Words cannot convey an adequate impression of its stupendousness. Charles Dickens, when he first visited America, felt himself unable to describe the scene, and only succeeded in eloquently sketching his emotions. "When we were seated in the little ferry-boat," he says, "and were crossing the swollen river immediately before both cataracts, I began to feel what it was; but I was in a manner stunned, and unable to comprehend the vastness of the scene. It was not until I came on Table Rock and looked—great heavens! on what a fall of bright green water—that it came upon me in

·its full might and majesty.   Niagara was at once stamped upon my heart, an image of beauty, to remain there changeless and indelible until its pulses cease to beat forever.   I think in every quiet season now, still do those waters roll and leap and roll and tumble all day long ; still are the rainbows spanning them a hundred feet below.   Still, when the sun is on them, do they shine and glow like molten gold.   Still, when the day is gloomy, do they fall like snow, or seem to crumble away like the front of a great chalk cliff, or roll down the rock like dense white smoke.   But always does the mighty stream appear to die as it comes down, and always from the unfathomable grave arises that tremendous ghost of spray and mist which is never laid.''

Niagara Falls are situated on the Niagara River, nearly midway between Lake Erie and Lake Ontario.   About a mile above the cataract the river is one continuation of rapids, which finally terminate in a perpendicular fall of 164 feet on the American side and 158 on the Canadian.   Goat Island, a quarter of a mile wide and half a mile in length, extends to the very brow of the precipice, dividing the falls into two portions, the higher of which is on the American side, while the greater width is on the Canada side.   The volume of water which constantly pours over this immense precipice and the power with which it sweeps everything before it is appalling.   No living thing has ever been known to go over and come out of the whirlpool below alive.   A pine board floated over soon comes to the surface in splinters.   About fifty years ago a vessel loaded with live animals went over the falls, and such was the eagerness of the sojourners at the hotels to see the sight that many of them jumped up from their dinners and·forgot to return to pay their bills.   The deep green color of the water and the effect in contrast with the white foam cannot be portrayed with pen, nor scarcely with painter's brush, though Church's famous picture in the Corcoran Gallery at Washington is marvellously near to it.   A beautiful bright rainbow rests above the water a few rods below the whirlpool, where the sun's rays are reflected through the mists, and is visible from morning till night in clear weather.   Considerable changes have taken place during recent years by the falling down of masses of rock, causing a slight recession of the cataract at some points.   Table Rock, once a striking feature on the Canadian side, has wholly disappeared. The chief points of interest to the visitor are Goat Island, reached by a bridge 360 feet long ; Luna Island, the Cave of the Winds, a spacious recess beneath and back of the American fall, and the Suspension Bridge below.   The old Terrapin Tower, which formerly stood out in the stream, 50 or 100 yards from Goat Island, from which a magnificent view was formerly obtained, became unsafe and was blown up with gunpowder in 1873.   Of late years the owners of property about the falls have fenced in all the approaches and points of interest for the purpose of exacting a charge for viewing them.   The Legislature of New York has been considering means of getting control of the property to make it a free park for all time, as it should be.   Niagara is reached from New York by the Erie and New York Central Railways, from Boston by the Boston and Albany, from Washington, Baltimore and Philadelphia by the Pennsylvania and Northern Central, and from the North and West via the Grand Trunk line. The hotel accommodations are ample and excellent.   It is a famous place for bridal couples, being equalled only by Washington in this respect.

There is one certain thing, says the author of ''Rambles of a Journalist,'' about Niagara ; it can have no rival.   Saratoga may become antiquated—the seashore a resort only for invalids. Fashions may change in regard to pleasure resorts.   Rival locations may compete by opposing attractions.   But Niagara can have no rival.   The flood will sweep on over the precipice, the waters will boil and foam, struggle and heave down the rapids, rushing on forever, and the roar of the cataract will be there forever.   In all the world there is but one Niagara, and all the world will visit the mighty show.   You may build up a city there, make long streets and ilne them with houses, and crowd them with people; and strip it of the things that nature spread out all around it; you may construct canals and erect machinery, but still the great cataract will be there, and the world will travel hundreds and thousands of miles to see it. They will go to the brow of the precipice to look down, and to the base of the precipice to look up. They will involve themselves in the mist and spray for the sake of gazing upon the rainbow that is above them. They will ramble on Goat Island by moonlight, listening to the roar of the waters, or enjoy its cool and pleasant shades at noonday. You may roll your great water-wheels in their ceaseless rounds; you may harness your machinery and set your great hammers in motion; your hundred strong hands may hurl the ponderous sledge against the ringing anvil; you may set your ten thousand puny machinists at pounding the iron and driving the spikes; make all the noise you can—and the roar of the cataract will drown it all.

The Caverns of Luray.

THE now famous Luray Cave in Virginia is a comparatively recent discovery, its existence having been first learned on the 13th of August, 1878, and not made publicly known until some time thereafter. The conical hill on the Newmarket pike, about a mile from the village

Virgin Font.

of Luray, in Page County, had long been known as Cave Hill, from the existence of a small cave near its summit; but the significance of certain sink-holes and standing ponds along its sides and about its base was not understood or suspected until a short time previous to the date above named. Mr. B. P. Stebbins, a photographer from Easton, Md., appeared in the locality, and, conceiving the view from surface indications, that a cave lay beneath the hill, induced several of the villagers to join him in the search for it. Together they went prospecting about the country, digging here and there, without success, until they were nicknamed "cave-hunters," and became the objects of good-natured ridicule. Their fellow-townsmen declared they were mistaking rabbits' holes for mare's nests and jumping rabbits for sprightly young colts. But finally the right hole was explored, and a depression in the hillside proved to be the entrance to the long-sought cave. One of the party—which now consisted only of Mr. Stebbins and Messrs. A. J. and Wm. B. Campbell—was lowered by means of a rope into the pit, and found himself in a narrow rift about fifteen feet long by five wide, with no apparent outlet. Closer examination disclosed a hole, through which, with some difficulty, he

Cathedral and Hall of Giants.

passed into a large open space, now known as Entrance Hall. Having abandoned the rope which connected him with his companions, he surveyed for some time with rapt interest the strange scene presented to his sight. The rest of the party becoming alarmed at his absence, another of their number was lowered in search of him. Together they returned to the upper world, and at night the party resumed their explorations with candles, getting as far as Muddy Lake—then a considerable body of water, since drained and replaced by a dry

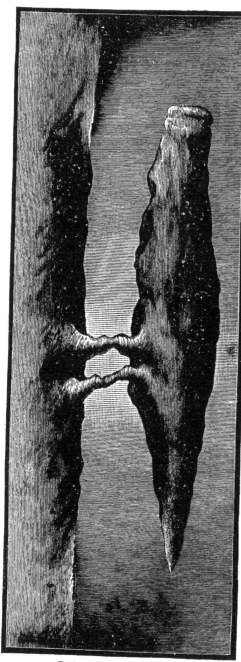

Curious Stalactite Growth.

cement walk—by which they were stopped and left in ignorance of the largest and grandest part of the cave. These, briefly condensed from Prof. Ammen's work, are the circumstances of the discovery of this great natural wonder. About two years later the property passed into the hands of the Luray Cave and Hotel Company, identical in interest with the Shenandoah Valley Railway Company, by whom numerous improvements have been made looking to the attraction and comfort of visitors. A handsome cottage has been built over the mouth of the cave, through which entrance is made, and in the interior cement walks, plank platforms, stairways, and railings have been provided wherever needed. The tallow candles formerly employed to illuminate the cave have been replaced with thirteen electric lights.

Entering the cave one is possessed of the feeling of having passed into a new state of being. Queer shapes present themselves at every turn, aping grotesquely the things of our past experience. Every object suggests some growth of animal or vegetable life, yet every resemblance proves illusive. There are glittering stalactites and fluted columns strong enough to bear a world; draperies in broad folds and a thousand tints; cascades of snow-white stone; and, beyond, a background of pitchy darkness in which the imagination locates more than the eye can see. But shortly the visitor begins to examine the objects more closely. First to attract attention is Washington's column, a fluted massive stalagmite about twenty-five feet in diameter by thirty in height, reaching from floor to ceiling. Stalactites depend on every side. From the centre of the roof one descends as aptly as if nature had designed it to support a chandelier. Passing on through Entrance Avenue there is seen a rounded bank of dip-stone, fringed beneath with semblances of dangling legs. Further on is the Flower Garden, a space inclosed with natural stalagmitic border and containing bulb-shaped stalagmites resembling bunches of asparagus, cauliflower, cabbages, etc., according to one's fancy. Beyond this is the Fish Market, where are distinctly seen hanging rows of fish—black bass, silver perch, mackerel, and, as the guide facetiously calls them, "rock" fish, and other varieties. The Smithsonian report says there is no difficulty about identifying the various species, some being gray all over, others having black backs and white bellies. The Elfin Ramble is a vast open plateau, estimated to be 500 feet in length by 300 in breadth. Still further on is Pluto's Chasm, the rift through which the god is supposed to have borne Proserpine to the under world. It yawns in a startling way, attaining a depth of seventy-five feet and a length of 500. At the bottom is the Spectre, a tall, white, fluted stalactite, covered about the upper part with a fringe of snowy draperies, and suggesting a meditative ghost.

But it would not be possible in this brief article to describe in detail all the fantastically curious formations within this cave. There is the Crystal Lake, the Virgin Font, the Frozen Fountain, the Organ, the Ball Room, and a hundred other varied shapes and places, all carved out of the blue limestone by the action of natural elements. Large areas, embracing some of the most wonderful parts of the cave, are not yet opened to the public. To quote still further from Prof. Ammen: "It is a task of recognized difficulty to describe the indescribable. This difficulty is enhanced, if possible, in the case of cave scenery, by the fact that the impressions it leaves upon the mind of the beholder differ, not so much in degree as in kind, from those of his past experience. A new order of sensations, ideas, and emotions demands a new vocabulary. The visitor who attempts a description must content himself, therefore, with seeking to impart his enthusiasm without hoping to fully trace its causes." The electric light heightens wonderfully the contrasts of light and shade, upon which cave scenery so much depends for its striking character. Under its glow the whiter formations shine with the lustre of pearl white, while the amber tints of the older and darker ones are changed for the color of gold. "There are," says one writer, "in these combinations of the picturesque with the statuesque, resemblances approaching at times the most advanced qualities of the sculptor's highest art. Indeed, it needs but a little play of the imagination to people these dusky chambers with conservatories rich with crystallized leaves and blossoms, with canopies of snow and ice, with crystal streamlets over which the glistening nymphs hum their peaceful tunes." It is impossible to estimate correctly the age of the cave or its formations. The rate of growth of cave formations varies with a score of circumstances, so that no general rule can be invariably applied. A tumbler standing five years under the drip of a stalactite was incrusted to the depth of an eighth of an inch. At this rate of growth, supposing all the conditions to be exceptionally favorable, a column one foot in diameter might be formed in two hundred and forty years. Under ordinary circumstances, however, it would perhaps require several thousands, some reckoners say tens of thousands of years. Dr. Porter, of Lafayette College, a distinguished scientist, in a recent lecture, quotes an eminent brother scientist as saying, concerning the Fallen Column, a gigantic formation weighing one hundred and seventy tons, that "four thousand years must have passed since its fall, and seven millions of years were consumed in its formation." This calculation is based upon the probable time which, in his opinion, it took to grow the vertical stalactites which have formed upon it as it lies.

Luray is on the line of the Shenandoah Valley Railroad. To reach it from New York, Philadelphia, and the North and West take the Pennsylvania Railroad. From Baltimore take the Western Maryland Railroad. From Cincinnati and the South and Southwest take the Chesapeake and Ohio Railway.

Watkins' Glen.

A FEW years ago but little was known of this picturesque and interesting resort beyond the confines of the county in which it is located. To-day it is renowned the world over for its wonderful scenery, and is annually visited by thousands of tourists, excursionists, and travellers from this and foreign countries. The village of Watkins, containing about three thousand inhabitants, is beautifully situated at the head or southern extremity of Seneca Lake, within the shadow of Glen Mountain. Seneca Lake is one of the most beautiful sheets of water in the world. It is entirely framed about with "precipitous, black, jagged rocks" and clean pebbly beaches. There is not a rod of swamp in its whole circumference, consequently it breeds no malaria, no mosquitoes. Excursions upon the lake to its many points of interest, in addition to the attractions of the Glen have made Watkins one of the most popular resorts in the country. The Glen is simply a vertical rift or gorge in a rocky bluff, some five or six hundred feet in height, through which rushes a mountain brook of purest water—now roaring and tumbling over rocks in foaming cascades, again plunging over ledges in beautiful falls, and anon eddying about in quiet little lakelets in the deep ravine, down upon which from high rugged crags or rustic little bridges the tourist may look and meet his or her face in the water. The Glen is divided into sections, each of which is given a distinctive name in accord with some one of its many beautiful or strange and wonderful features. The division at the entrance, which is only a quarter of a mile from the station, is named Glen Alpha, and the section at the terminus, about

Mammoth Gorge.

three miles above, is called Glen Omega. A short distance above the entrance to Glen Alpha, a narrow but safe bridge crosses the chasm, from which an excellent view is obtained of Minnehaha Falls, one of the prettiest cascades in the Glen. Farther up, at a point where the high and rugged walls draw close together, is 'Cavern Cascade,' where the water falls over the rocks into a gloomy basin. The tourist has, for some distance up to this point been traversing a narrow footpath cut out of the face of the cliff. He now leaves this path and climbs the long staircase which crosses the chasm, and ascends for fifty feet at an angle of ninety degrees to another footpath on the other side of the Glen. From this point the path leads around moss-covered boulders, along steep rocky slopes and ledges, up a succession of stairways crossing from side to side, until by an ever upward climbing other pathways are gained, pursuing which the stairway is reached which leads to the Mountain House, perched on a shelf quite overhanging the gorge. The gloomy division beginning at this point is called Glen Obscura, passing through which and by the Sylvan Rapids, across a bridge to the other side of the Glen the narrow gorge expands into an enormous amphitheatre, to which has been given the name of Glen Cathedral. Of the many

remarkable chambers the Cathedral is, perhaps, the most imposing. It is an immense arena a thousand feet long with walls of solid rock rising perpendicularly to a height of three hundred feet, while the floor is almost as level as if it had been paved by human hands. Into this mighty chasm the waters spring with a frightful headlong leap, bathing the sides with feathery spray, then quietly spreading over the rocky floor form the lovely pool of the Nymphs. From the north side of the Cathedral, the Grand Staircase, thrown across the ravine to a higher shelf of the cliffs, leads to the Glen of Pools. Beyond the Glen of Pools is the Giant's Gorge, at the upper end of which Rainbow Falls, one of the most interesting and beautiful features of the Glen, is reached. The path passes behind the fall and leads up another stairway to the Shadow Gorge, at the head of which, by a pathway cut in the sides of the rugged rocks, Pluto Falls are reached. Here the waters pour down from a rocky parapet into a deep, dark basin. The especial points of interest between this spot and Glen Omega, are Glen Arcadia, Arcadian Falls, Elfin Gorge, Glen Facility,

Artist's Dream.

Glen Horicon, Glen Elysium, and Omega Falls. Besides the points and objects mentioned, there are a hundred others in this marvellous Glen, each possessing particular features of interest—spots where hours may be spent in watching the restless waters pouring down from rocky heights, leaping over huge boulders, or sweeping across smooth beds of shining pebble. The atmosphere in the Glen, even in the hottest day, is cool and moist.

Watkins is reached from Baltimore *via* the Northern Central Railroad ; from New York and the West by the Erie Railroad to Elmira, and thence *via* the Northern Central, or by the New York Central to Seneca Lake, and thence by steamer to the village.

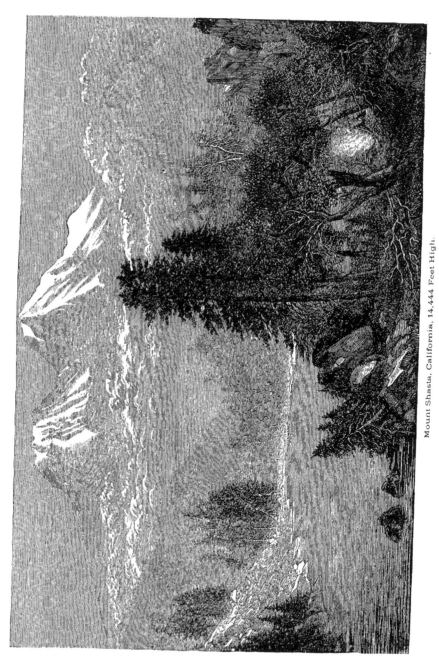

Mount Shasta, California, 14,444 Feet High.

(56)

Mountain Resorts.

"Here mountain on mountain exultingly throws,
 Through storm, mist, and snow, its bleak crags to the sky;
 In their shadows the sweets of the valley repose,
 While streams, gay with verdure and sunshine, steal by."

OUNTAINS are symbols of grandeur and sublimity. They have been called "God's eternal sentinels," because more than all else in nature they bring man to a contemplation of his own littleness and the awful extent of infinite power. No man can behold the aspects of a nobly-uplifted pinnacle or dome without realizing that his thought is expanded, unchained and newly-gifted. From the earliest dates in the world's chronology mountains have commanded the supremest worship and admiration, and profoundly symbolized noted epochs in the panorama of history and events. To use the words of Professor Winchell: "There is more in mountains than the novelty of the outlook from their summits. They stir the higher susceptibilities of the intellect by their magnitude, their loftiness, their grandeur, and the rugged unapproachableness of their peaks." They fire the soul with a spirit of veneration—they are the symbols of eternity and boundless power. They are the homes of frost, and silence, and mystery—the brows which bear the wreath of the clouds—the eyries of the lightning and the thunder—the palaces of infinite greatness and majesty. Every lover of nature is a lover of the mountains, and every student of science and natural wonders finds a workshop and a study amid their rocks and crevices. The botanist finds there his rarest flowers and plants, and the geologist his most valuable specimens. The pleasure traveller and health-seeker find in the mountains the rarest air, the sublimest scenery, the most enjoyable exercise, and in many cases the greatest benefits. Those who have lived among mountains are seldom contented elsewhere, and those who once spend a vacation in them look eagerly forward to another.

He who first met the highlands swelling blue,
 Will love each peak that shows a kindred hue;
 Hail in each crag a friend's familiar face,
 And clasp the mountain in his mind's embrace."

The mountains of our own land embrace every degree, from the "green hills" of Vermont to the picturesque Catskills, the wild Adirondacks and Alleghanies, the beautiful Blue Ridge and the lofty snow-capped Rockies and Sierras. While no writer can ever hope to poetically create another Ararat, Sinai, Calvary, Pindus, Olympus or Parnassus, the time will certainly come when the fame and influence of our noted earth-giants, with their incomparable forests, and waterfalls, and domes, and lakes, will outrank and eclipse even that of the Alps, the Apennines, the Cevennes, the Vosges, and the Cote d'Or.

Colorado.

OWING to the extent and grandeur of its mountain scenery, Colorado ranks first among the mountain regions of our country. The whole State is one vast summer resort, or tourists' home, and the stream of sight-seers, pleasure and health-seekers, which annually flows into it, grows larger year by year. This region has been frequently called the "Switzerland of America;" but there are so many localities to which this term has been applied, that it scarcely conveys its full meaning. By the concurrent testimony of travellers, the scenery of the Rocky Mountains is not inferior to that of the world-famed Alpine region in Europe. Yet there are points of difference, chiefly in the surpassing magnitude and grandeur of these

5   (57)

Chicago Lake—The Highest in North America.

immense Rockies and the wonderful cañons among them, whose unique and even fantastic formations are unequalled anywhere in the universe. These river cañons or deeply-cut ravines, that are found in all the more elevated portions of Colorado, constitute a peculiar and striking feature of the great Rocky Mountain system. In the countless ages of the past, the waters of the streams have worn channels deep down into the hearts of the mountains, leaving the perpendicular granite or sandstone standing on either side for hundreds, and in some localities thousands, of feet. Nowhere are the grand and wonderful in nature more effectually illustrated than in these mountains and cañons. There are no less than fifteen peaks in the State, each with an altitude but little below that of Mt. Blanc; and, in extent of surface, one of these great peaks exceeds the entire area of Switzerland. To gain some idea of the extent of Colorado scenery, let it be remembered that the State is larger than Great Britain, comprising an area of 67,420,000 acres, of which one-third only is grazing or agricultural territory, while the remainder is the vast upheaval known as the Rocky Mountains, the "back-bone of the continent," describing a tortuous course north and south through the State, which "covers more outdoors" than any other State in the Union except California and Texas. This vast area lies between the thirty-seventh and forty-first parallels of north latitude, and the one hundred and second and one hundred and ninth meridians of west longitude. Its average extent, north and south, is 275 miles, and east and west 380 miles, the total area being 104,500 miles.

Approaching Colorado from the east, the traveller makes a gradual ascent after leaving the Missouri River, and the eastern border of the State is crossed at an elevation of 4090 feet. In the central part of Colorado the mountains form four vast basins, called parks,—North Park, South Park, Middle Park, and San Luis Park. North Park, with its area of 2500 square miles, at an elevation of about 9000 feet, has a north-central location. Just south of North Park is Middle Park, with its area of 3000 square miles, at an elevation of 8500 feet. Still

Winnie's Grotto—Walls 2000 feet high.

south of Middle Park is South Park, with its area of 2200 square miles, at an elevation of 9500 feet. The fourth Park, San Luis, is near the south line of the State, has an area of 8000 square miles and an elevation of 7000 feet. In these parks are numerous small lakes, besides many beautiful streams and mineral springs, which are becoming popular resorts. The now famous Twin Lakes in Middle Park are, with a single exception, the highest bodies of water in North America. Some of the numerous summer residents of the locality have provided

Running the Rapids, Colorado River.

themselves with sail-boats, and enjoy the novelty of yachting at an elevation of 11,000 feet. Hunting and fishing have also been bountifully indulged in by tourists fond of these sports. Game was, a few years ago, very plentiful, especially in North Park, which was the natural herding-ground of thousands of elk, antelope, deer, and mountain-sheep; but their numbers are becoming considerably diminished, though the pursuit is still sufficiently rewarded to give zest to the sport. The Earl of Dunraven, in his very fascinating account of a hunting season in Colorado, thus sums up his impressions: "In spring and summer the scenery and climate are very different. Ice and snow and withered grass have passed away, and everything is basking and glowing under a blazing sun, hot, but always tempered with a cool breeze. Waterfowl frequent the lakes, the whole earth is green, and the margins of the streams are luxuriant with a profuse growth of wild flowers and rich herbage. The air is scented with the sweet-smelling sap of the pines, whose branches welcome many feathered visitors from southern climes; an occasional humming-bird whirrs among the shrubs, trout leap in the creeks, and all nature is active and exuberant with life."

The mountains of Colorado are drained chiefly by the Rio Grande, the Arkansas and the

Platte rivers. The latter runs through the valley in which Denver is situated ; and, though this is said to be a country where rain seldom falls, and where agriculture is only possible by irrigation, it has several times gone on the rampage and caused great damage. The glories of Platte Cañon and the Grand Cañon of the Arkansas have been most written about, but the walls of the Colorado and Gunnison rivers, in the western part of the State, are far more massive and wonderful. In many sections they rise without a break or an incline to heights

of thousands of feet, and along the Colorado continue in that way with hardly an outlet of any kind for hundreds of miles. Major Powell, of the United States Geological Survey, gives, in the report of his explorations of this river, the only graphic account of its wonders ever printed. The Grand Cañon of the Gunnison is another of the world's wonders. Its walls on either side of the stream, and bordering it for miles, are usually not far from 300 feet in width and are composed of stratified rock. In places their perpendicular sides, rising from the water from one to three thousand feet, terminate in level summits surmounted by a second wall of prodigious height, thus forming a cañon within a cañon. Through the chasm between these giant formations and huge bastions and turrets, one above another, dashes the river, its surface white with foam.

Outside of Denver, usually the first point visited by all tourists, the chief places

Swallow Cave, Colorado River.

of interest and the ones most convenient and accessible, are : Colorado Springs, Manitou and surroundings, Boulder Cañon, Greeley, Idaho Springs, Georgetown and vicinity, Central City, Pagosa Springs and the Parks. Of course the mountain scenery is everywhere, and mining operations, of interest to many, are to be seen in every part of the State. The beautiful city of Denver, with its progressive spirit and metropolitan appearance, is doubly attractive after the long journey across the plains, and the fascination of first sight is increased on closer acquaintance. There is a dash and animation about the place, with a finish and elegance that suggest prosperity, wealth and stability quite as much as the aggressive frontier. Denver is the best built city between St. Louis and San Francisco, and its growth at the present time is more rapid, and its prospects more brilliant, than any other city in the whole country. Its population is now about 70,000. The city is built mainly on ground sloping slightly towards the mountains, which rise so grandly along the entire western horizon, the line of vision taking in the "snowy range" and its outlying foot-hills for a distance of one hundred and

Mary's Veil, Upper Falls on Pine Creek.

fifty miles. The streets are broad, solid and cleanly, and are lined in all directions by massive blocks, elegant residences, green lawns and handsome shade trees. The city is well provided with hotels, most of them first-class. The Windsor, the newest, largest and best, is equal to any in the whole country. Except for a few of the hottest days, Denver is a delightful place during the entire summer. Those who prefer a more rural retreat for a portion of the time go out to the Springs or to Greeley. At the latter place excellent accommodations are found at the Oasis Hotel. Nearly all tourists make Denver headquarters and plan their trips in different directions from that point. The next place in importance, especially from the tourist's standpoint, is Colorado Springs, the most beautifully located, cleanest and cosiest appearing place in the State. Near here is Manitou, known as the "Saratoga of the West," and within a radius of five or six miles are some of the most interesting features of Colorado scenery. At Manitou, which is situated just at the opening of the Pike's Peak trail, are located the most famous mineral springs in this region. The waters are strongly charged with carbonic acid and contain carbonates of soda, lime and magnesia in various proportions. Broad claims are made for the medicinal properties of these waters, the opinions of professors of chemistry being quoted to the effect that they excel the "Ems" and the "Spa," two of the most famous groups in Europe. The elevation of this locality is higher than that of Denver, or a little over 6700 feet. There are splendid drives in all directions, and within a radius of seven or eight miles are numerous attractions and points of special interest, including the Garden of the Gods, Glen Eyrie, Ute Pass, and Monument Park. The formations from which the latter takes its name are among the greatest curiosities to be seen in Colorado. Pen cannot well describe them. They consist of a series of curiously-shaped natural monuments which have been formed from sandstone rock solely by the action of the elements, a thin stratum of iron on the top having protected these particular pieces and preserved them. No accurate estimate can be made of the hundreds of years this work of nature has been in progress. There are perhaps two hundred of the peculiar formations of different sizes and shapes, some of which are really fantastic, the whole covering an area of five hundred acres, in the midst of a perfect natural park. The Garden of the Gods is also a remarkable freak of nature, partaking somewhat more of the grand and imposing. It is a secluded spot, hemmed in by great rocks stood up on edge and on end. They are some of the more marked of the numerous evidences on every hand here of a grand upheaval some time in the past. These tremendous copper-colored slabs looming up, some of them 350 feet high, are an imposing sight. Some look like enormous pillars; others are cathedral-shaped towers, the whole forming a scene at once weird and enchanting. The tourist in search of either health or pleasure may profitably spend many days or even weeks in the vicinity of Colorado Springs and Manitou. The two points are only five miles apart and connected by a narrow-gauge railroad, by which the fare is twenty-five cents for a round trip. The air here is bracing, and there is ample amusement for the lovers of nature in the cañons, grottos, mountains, and passes. First of all, there is that giant sentinel, Pike's Peak, towering over plain and foot-hill, the view from whose summit is indescribably grand. Although this attains the enormous altitude of 14,147 feet, by following the trail it can be ascended on horseback. On the barren, rocky mountain-top is a government signal-service station. To witness sunrise from this elevated position is an experience long to be remembered, as is the whole day's trip, for it is a laborious and tiresome journey. The spectacle of a snow-squall on this Peak in midsummer is a treat, and may be often witnessed from the Garden of the Gods and other points in the range of vision. Though the most famous Peak in Colorado, and seen at the greatest distance in all directions, Pike's is not the highest, Gray's

Island Monument, Glen Canyon.

Peak, twelve miles above Georgetown, being 200 feet higher.

The most entertaining tour to be made in Colorado, and the one embracing the greatest amount and variety of scenery for the time and expenditure required, is from Denver by the Colorado Central Railway, now a part of the Union Pacific system, through Clear Creek Cañon to Georgetown, Idaho Springs, and Central City. These points with their surroundings furnish material for weeks of pleasant exploration, or they may be hastily seen in two

days. Picturesque Clear Creek Cañon has been often portrayed, but it must be seen to be appreciated. Passing through it is almost like going into an immense cave. Its towering peaks and overhanging rocks are high above on either side, sometimes shooting straight up, with walls as perpendicular as those of a cavern, and almost shutting out the light of day. To stand on the rear platform of the train affords a grand sight, and to see the panting little iron horse twisting around in the crevices of the rocks, as it were, often apparently turning around to come back at you, is a most novel and exhilarating railroad experience. Idaho Springs are reached soon after emerging from the cañon. Though not as noted in the world of fashion as Manitou, these springs are probably the best in Colorado, and the air of the locality is a perfect tonic, unequalled anywhere. A few miles distant are those noted resorts, Chief Mountain and Chicago Lake, the latter being the highest body of

Cathedral Rock, Garden of the Gods.

still water in North America. The drive over the mountain to Central City, a distance of six miles, is a most enjoyable trip.

Sixteen miles beyond Idaho Springs, situated almost in the heart of the mountains, fifty-two miles from Denver, is Georgetown. It is not only picturesque in appearance, but unique, and will strike the new-comer from the East as wholly unlike anything he has ever seen before. All around are curiosities and places of interest. There are drives and walks unsurpassed, with lakes, and mines to visit and mountain peaks to climb. In the sides of the steep mountain around and above the town, tunnels and shafts without number have been dug in the eager search for gold and silver, both of which have been obtained in the locality. Gray's Peak, one of the four highest in the whole range, is only twelve miles distant. Tourists usually make the trip to it on horseback, and those who wish to

Devil's Gate, Vicinity of Georgetown.

enjoy it to the best advantage, and to save themselves unnecessary fatigue, take a part of two days for it, spending the night at a cottage at the foot of the mountain. By this means the ascent can be made in early morning, always the best time. It is a hard climb up the narrow winding trail, where to look back makes one's head swim, and where a misstep or a stumble would precipitate horse and rider down the terrible rocky incline to almost certain destruction. But the magnificence of the scene repays many times over the labor and risk of reaching it. This cold stony summit points up through the clouds 14,351 feet above the sea level, and in the hottest days of August one requires extra wraps while standing upon it. Looking down hundreds of feet below may be seen immense snow-banks which the summer's sun has failed to dissolve. Resting over the tops of lower mountains are seen great white clouds, which from above, with the sunlight shining on them, also look like sheets of snow. For hundreds of miles in every direction mountain peak after mountain peak meets the view, snow-capped and rock-bound, "grand, gloomy and peculiar." In the clear rarefied atmosphere there is almost no limit to one's vision with a good field-glass. Denver lies 60 miles to the south. Pike's Peak, 150 miles distant, appears to be scarcely a gunshot away. South Park seems to be almost at your feet, while ranges of moun-

Sandstone Formations, Monument Park.

tains in Wyoming and New Mexico are plainly visible. One of the most noted mountains of Colorado—the Mount of the Holy Cross, so called because of the cross-shaped snow-lines always visible near its summit—is also readily seen. Victor Hugo tells us that "every condition has its instinct," and he who finds himself for the first time face to face with the Rocky Mountains has an appalling sense that he has not only overrated his individual importance in nature's economy, but has likewise undervalued the influence of inarticulate nature upon himself. Nothing can transcend the majesty of these snow-capped mountains! You gaze upon them in mute wonder until you grow abstracted and out of self into the idea of perpetual greatness. You do not think—only feel—and somehow the Eastern world that you have left behind, with its glitter and gloom, its envious struggles and manifold defects, fades into insignificance in view of this endless range of divine architecture, and you are for once an humble worshipper at the pure shrine of sublimity.

Three miles up a mountain gulch above Georgetown is what is known as Green Lake. It is a large basin, at an elevation of 11,000 feet, filled with water 75 feet deep, almost as cold as ice, and of a greenish hue. The lake is half a mile long by a quarter of a mile in width, and overlooked on all sides by an impregnable mountain wall. The water is at times very clear and transparent, and in one portion of the lake is what the natives call a "petrified forest." The tops and dead branches of standing trees are distinctly seen, though seventy-five feet below the surface. In this lake the propagation of fish is extensively carried on, and the water literally swarms with beautiful trout and salmon. They are so tame that droves of them will come close to the shore and eat crumbs from visitors' hands. There are usually about 20,000 trout in the lake and several thousand young ones in the tanks below. To feed them requires fifty to sixty pounds of ground meat per day in summer, and a larger quantity in winter. No fishing is allowed, and an armed man patrols the bank at night to prevent the destruction of these pets. The object of this enterprise is to stock the mountain streams for food and sport. All the waters of Colorado are being rapidly depleted of their beautiful finny inhabitants, and fishing there is not what it was a few years ago. Over the mountain, about twenty miles from this locality, wedged in two ravines, the tourist will find the towns of Central City and Nevaville. It is worth the trouble of getting there just to see them. The first sight of these cities of the hills is one not soon to be forgotten. There is a novelty in the scene which attracts in spite of the general barrenness of the landscape, the forest having long since been consumed in furnaces and mines. Thus the numberless prospect holes, dump piles, shaft cuts, and tunnels, that scar the earth's surface, are all the more plainly visible. Streets and houses

Gray's Peak.

stand almost in tiers one above the other in narrow ravines and gulches. The towns centre where two streams and gulches unite, and the main thoroughfare, over three miles in length, winds through and around granite hills. Far up the giddy slopes, on either side, hang cottages and mine buildings, seemingly ready to topple one on another.

An excursion to Middle Park can be made from Georgetown in two days, and many tourists, who have the time to spare, avail themselves of it. A stay of any length in the park is best enjoyed by "camping out." It is a region best suited to "roughing it," and the attractions are largely such as invite sportsmen and others inclined to that sort of life. For those who expect to spend several weeks in Colorado, this method of life is very desirable, not only in the parks but in other sections; and, while having its advantages in many respects,

Mount of the Holy Cross.

is not more expensive, including the entire outfit, than living at hotels. Middle Park, while not a paradise, as often represented in railway and other publications, possesses many attractions. The broad expanse of mountain scenery, unfolded from the passes of the Sierras or the valleys of the park, and the rolling prairies and river bottoms, with their luxuriant carpeting of grasses and flowers, diversified with groves of pine and aspen, form a picture but little short of enchanting. There is everything that goes to make a mountain ramble enjoyable,—cool, invigorating atmosphere, bright skies, good hunting and fishing, mineral waters, clear lakes, translucent streams and sparkling waterfalls. Once over the great Divide whose lowest passageway is more than two miles above the sea, one can revel in the unrestricted freedom of

mountain life in one of nature's favored localities. Among the places not heretofore mentioned are the Pagosa Springs, which lie four miles south of the San Juan range, on the river of the same name. The chief attraction is a cluster of hot springs, the largest of which is forty feet in diameter, the water being exceedingly hot and charged with saline material. The celebrated Poncho Springs are located a short distance from South Arkansas, and are fifty in number. The locality offers numerous attractions as a pleasure resort ; the scenery is grand and inspiring, views being had of Mounts Ouray, Shawano, Antero, Harvard, and Princeton. In the Wet Mountain valley, which is an old lake basin, lying between the Sangre de Cristo range and the Greenhorn Mountains, are three prosperous towns, Silver Cliff, West Cliff, and Rosita. Professor Hayden regards the view of the Sangre de Cristo range, from the Wet Mountain valley, as the grandest in Colorado. In this portion of the range rise four peaks, all of which are higher than Pike's. Saguache, thirty-three miles from Del Norte, is located near San Luis Lakes,

Camping at Flaming Gorge.

a large body of marshy land and shallow ponds, in which ducks are found in plenty.

Tourists may expect to encounter many interesting and almost irreconcilable freaks of nature during extended rambles in the Rocky Mountains. A writer of some note thus speaks of personal observations: "While crossing the 'range' which girdles North Park, one July day several summers ago, we were among snowbanks much of the time, and at night our camp was made by a great bank of glittering 'beautiful,' on account of the abundance of water, fuel and horse-feed in that vicinity. The bank was higher than our heads, and slowly melting under the influence of the July sun. At the very edge of the snowy mound we found strawberries in full bloom, and within ten feet could be counted half a dozen varieties of flowers. Water froze hard in camp utensils during the night, and the customary white frost was everywhere visible in the morning. Morning after morning, in our wanderings at these high altitudes, have we shaken the crisp scales of white frost from our blankets and looked around upon a scene of apparent desolation. Brilliant flowers of the evening before wilted into ruins, and the splendid tall blue grass that looked a delicious morsel for stock at sunset, was bent, and sometimes broken, as with the weight of a night's winter. But an hour of sunshine always changed the scene to one of springtime freshness, and often the flora, apparently most delicate, rallied first under the magic influence." An experience to be remembered by every tourist who meets with it, is getting caught in a storm up in the mountains. The rain-clouds do not overspread the whole heavens as in the Atlantic States, but pass over in areas of narrow width, following up the mountain spurs and chains, and often, when the rainfall on a mountain-top or mountain-side is sufficient to transform the tiny rivulet or brooklet into a raging torrent of water, there will be in the valley below, only a mile or two distant, continued sunshine and a balmy and fragrant atmosphere. It is a grand and glorious sight to witness a thunderstorm in these mountains, if you only happen to be at a safe distance. Then, too, listen to the rolling, almost deafening reverberations as the thunder-cloud passes over some lofty peak or range, and to witness the vivid play of the forked lightning as it flashes from cloud to cloud, or darts meteor-like from crag to crag, while you are basking in the beautiful sunshine, is glorious in the extreme. But to happen to be in the path of this rapidly moving storm is to get such a

Green Lake.

drenching as one may never forget. During the month of August these storms occur in the mountains almost every afternoon, between one and four o'clock. They come without more than a moment's warning, and there is no time for getting away from them.

Concerning the climate of Colorado various impressions prevail and much misunderstanding exists as to the effect of it upon different organizations. It is undoubtedly variable in some respects, but two things can always be depended upon in the summer season—pure air and plenty of sunshine. As a health resort the locality cannot be recommended indiscriminately for all sorts of people, with all sorts of diseases, as was done by interested parties a few years ago. To those in the enjoyment of ordinary health the sensations experienced in crossing the ascending elevations of the great plains, and in the higher altitudes at the base of and within the mountains, are in a notable degree pleasant. The dryness and rarity of the atmosphere, together with its remarkable electrical effects, combined with other peculiarities of the climate, excite the nervous system to a high degree of tension. Among the diseases which a visit to the Rocky Mountains will generally relieve, and often cure, are : Asthma, the earlier indications of pulmonary consumption, chronic bronchitis, certain forms of dyspepsia, and malarial poison. But it has been demonstrated that persons in the later stages of consumption go to Colorado only, in many cases, to die. Of the hundreds of patients of this class who have sought these high altitudes in the past some have found health, while many have sooner or later retraced their route in rapid decline. Any such, hoping for a cure, must not postpone too long the day of starting. And all persons in ill health are warned against making the transition from the lower to high altitudes too suddenly. It is always best to make one or two stops between Kansas City and the mountains.

The drawbacks to a Colorado tour are the same as are encountered in all these long journeys to the great West, though they do not exist in the same degree here as in the trip to Yellowstone Park. There is no staging to get to Colorado, and there is, as a rule, no lack of accommodations after arrival, especially in the centres. There are now four routes across the plains to Denver from Chicago, viz. : the Rock Island and Union Pacific, *via* Omaha ; the Rock Island and Atchison, Topeka and Santa Fé, *via* Atchison and Pueblo ; the Chicago, Burlington and Quincy, and the Rock Island and Kansas Pacific (now Kansas branch of the Union Pacific), *via* Kansas City. The latter is the shortest and most desirable route. In Colorado railway fares are enormously high—generally ten cents per mile—but competition may in time bring them down. The fare between Denver and Pueblo, 118 miles, has always been $10 by the Denver and Rio Grande, but the opening of the new Denver and New Orleans road caused a war of rates, and last year passengers were carried between the two cities for $1.

The hotel accommodations throughout Colorado may be termed "fair to middling." In Denver the Windsor is equal to the best in New York or Chicago, as is also the Oasis at Greeley. At Colorado Springs, Georgetown, Idaho Springs and Central City the average rate is $3 per day. The rates of the Manitou hotels are $4 and $5 per day. For all tourists to the mountains camping out, with a "Burro" pony to ride from point to point, is the least expensive and most satisfactory arrangement. These ponies can be purchased, with a complete outfit, for $50, and sold after use at a small sacrifice.

California.

BUT for its great distance from the populous portions of the East, California would probably be the most frequented, as it is the most attractive State in the Union. Though it has been celebrated in books, newspapers, and magazines for twenty years, it is really but little better known to the great mass of tourists than it was to Swift when he wrote his description of the flying island of Laputa. "There have been Americans," says Charles Nordhoff in his excellent work on California, "who saw Rome before they saw Niagara ; and for one who has visited the Yosemite a hundred will tell you about the Alps, and a thousand about Paris. But I would like to induce Americans, when they contemplate a journey for health, pleasure, or instruction, or all three, to think of their own country, and particularly of California. There, and only there, on this planet the traveller and resident may enjoy the delights of the tropics without their penalties ; a mild climate, not enervating, but healthful and health-restoring ; a wonderfully and variously productive soil without tropical malaria ; the grandest scenery, with perfect security and comfort in travelling arrangements ; strange customs, but neither lawlessness nor semi-barbarism." This is a glowing picture, but it is not overdrawn. It is undoubtedly true that California has a climate unequalled in any other part of our country, and that the scenery of her mountains and the Yosemite Valley ranks among the greatest wonders of the New World.

To most Eastern people California is still a land of big beets and pumpkins, of rough miners, of pistols, bowie-knives, abundant fruit, queer wines, and high prices—full of discomforts, and abounding in dangers to the peaceful traveller. But the tourist of to-day finds that the conditions of '49 have passed away—that California is thoroughly civilized, abounding in comforts, luxuries and endless delights. After spending a few days in San Francisco looking at the strange sights in the streets and visiting the Cliff House to get a view of the harbor, the first place next visited is the world-famed Yosemite Valley. Of this marvellous valley, where the most exquisite pencillings of nature have fulfilled matchless conceptions, an enthusiastic writer has said : "Yosemite conveys to the soul of man, through the eye, what might the orchestra of Heaven, through the ear, were peals of thunder compassed into harmonious notes of music, then suddenly silenced, and followed amid instant stillness by nature's most tiny voice." Another, who had written extensively of the scenes met with in a tour around the world, upon taking his first view from "Inspiration Point," said: "Like a spendthrift in words, the only terms applicable to this spot I have wasted on minor scenes." All writers agree that language fails to adequately express the emotions felt or convey the impressions obtained upon a first visit. Standing upon "Inspiration Point," the tourist obtains the first and most impressive view of the valley, and one that will remain ineffaceably stamped upon his memory. After satisfying the senses with one rapid, general survey of the valley, the eye rests involuntarily upon "El Capitan," the monarch of rocks, and the most matchless piece of natural masonry in the world; then the vision wanders to the opposite side, and takes in the beautiful waterfall known as the "Bridal Veil;" then the "Cathedral Rock;" then, back again, on the left, to

Bridal Veil Falls, Yosemite Valley.

the "Three Brothers," and, in the distance, the "Dome," "Half Dome," and many other masses of perpendicular granite walls, majestically lifting themselves to the sapphire heavens. The valley, which is some six miles in length by less than a mile in average width, is about 4000 feet above the level of the sea, and is thickly wooded and scattered all over with floral offerings, rich and varied, and abundant beyond the gardens of wealth and taste. And, amid the transcendent grandeur of the valley, meanders a stream as cool and crystal-like as the upper fields of imperishable snow and ice from which it takes its source. On the crest of the mountains, and at their base, and all along the mountain trails, "gush frequent springs for the thirst of the traveller, shooting their sparkling rills across his path as soon as his lips are parched, and inviting him to stoop and drink of a nectar cool with dissolving snows.''

The most attractive and beautiful object in the Valley, from March until July, is said by Major Truman, in his recent guide-book, to be the Yosemite Falls. The name is Indian and signifies large grizzly bear. These Falls are divided into three sections,—first a perpendicular descent of 1500 feet, then 600 feet of cataracts down a shelving ledge, and then a final leap of 400 feet. Professor Whitney concludes a description of them as follows: " As the various portions of the Falls are nearly in one vertical plane, the effect of the whole is nearly as grand, and perhaps even more picturesque, than it would be if the descent were made in one leap from the top of the cliff to the level of the Valley. Nor is the grandeur or beauty of the fall perceptibly diminished by even a very considerable diminution of the quantity of water from its highest stage. One of the most striking features of the Yosemite Falls is the vibration of

Three Brothers, Yosemite Valley.

the upper portion from one side to the other, under the varying pressure of the wind, which acts with immense force on so long a column. The descending mass of water is too great to allow of its being entirely broken up into spray; but it widens out very much towards the bottom—probably as much as 300 feet, at high water, the space through which it moves being fully three times as wide. This vibratory motion of the Yosemite and Bridal Veil Falls is something peculiar, and not observed in any others, so far as known; the effect of it is indescribably grand, especially under the magical illumination of the full moon.'' The gem of the Valley is Mirror Lake, which, in order to see the reflections, must be visited early in the day. Major Truman thus describes a recent view of it: " We shall never forget the last time we visited this lovely spot. Neither the glowing harmony of Byron nor the exquisite pencil of Raphael could have adequately delineated the incomparable splendor of that radiant scene. The sapphire heavens were untouched by atmospheric speck, and there was an ineffable beatitude in the deliciousness of the air. The Half Dome, with its storm-written hieroglyphics, stood above us in the sky and beneath us in the water, and we watched impatiently for the appearance of the imperial orb which had really dazzled us from our comfortable beds two hours before. At half past six o'clock a marvellous maze of opalescent cirra came suddenly over the summit, and chased each other rapidly across the silent lake; then followed processions of cumuli in pink, purple, crimson, violet, emerald, orange and dun; and then came the king of day in gorgeous state; and we gazed at it for some time in the waters as it flung its way triumphantly across its magnificently-frescoed track." But the attractions of this Valley are too numerous to be set forth in detail here. Perhaps the best known, and ranking with Mirror Lake among the most beautiful objects it contains, is the Bridal Veil Fall. To obtain an idea of it fancy a sheet of milk-white foam, seventy feet across, falling with a slight outward curve one thousand feet sheer descent, shattered into

Lake Tahoe.

spray near the foot and on the sides, which is blown about by the wind, and thrown back by the rebound till the base of the fall is quite hidden—then imagine the sun shining through this boiling mass of foam and mist, and watch the rainbows spanning the stream in concentric circles, as vivid as strips of brilliant ribbon, rainbows on each side, broken rainbows quivering down and others rising to meet them, every neighboring bush crowned with rainbows, and even the turf for rods around glowing with the richest colors, and all these shifting, changing, blazing, fading and forming again. Professor Whitney says of it: "The effect of the fall, as everywhere seen from the Valley, is as if it were 900 feet in vertical height, its base being concealed by the trees which surround it. The quantity of water in the Bridal Veil Fall varies greatly with the season. In May and June the amount is generally at the maximum, and it gradually decreases as the summer advances. The effect, however, is finest when the body of water is not too heavy, since then the swaying from side to side, and the waving under the varying pressure of the wind as it strikes the long column of water, is more marked. As seen from a distance at such times, it seems to flutter like a white veil, producing an indescribably beautiful effect. The name 'Bridal Veil' is poetical, but fairly appropriate. The stream which supplies this fall, at the highest stage of water, divides at the base into a dozen streamlets, several of which are only just fordable on horseback.'' Merced River is a pretty stream, which takes its source from the snows and lakes of the high Sierra, and dashes down into the Valley from innumerable cascades and waterfalls. Its banks are adorned by pine, fir, alder, spruce, poplar, and manzansta, and during the spring and summer months

with myriads of flowering plants and shrubs. During the months of May, June, and July, in particular, the California lilac, mariposa, azalea, and an infinite variety of smaller wild flowers are in full bloom and perfection, displaying all the rich colors of an Axminster, and which, interwoven with the emerald groves which enliven the banks of the Merced, constitute a piece of mosaic unrivalled in nature or art. The balsamic odors which escape the pines and firs add spice to the fragranee of the azalea and lilac, which freight the atmosphere with their aromatic sweets.

The Yosemite Valley is situated about 150 miles in an almost easterly direction from San Francisco, and nearly midway of the State from north to south. It was for many years the rendezvous or permanent abiding-place of hostile Indians, who had a legend for every point of interest, whether water or rock. The place was first seen in 1850

Rounding Cape Horn.
FROM NORDHOFF'S CALIFORNIA, PUBLISHED BY HARPER & BROS.

by a number of white men who had formed themselves into a military company to punish or compel peace with bands of murderous Indians. An expedition under Captain Boling invaded the aboriginal stronghold and obtained possession, only to be in turn annihilated some time later. After peace had been secured the Valley was occasionally visited by plucky tourists who had heard of its wonders from the soldiers. In 1855 J. M. Hutchings, publisher of the California Magazine, being engaged in gathering materials for the illustration of California scenery, organized an expedition which really made the first party of tourists to visit the Valley, and which makes Mr. Hutchings's name inseparably connected with it. During the year 1856 a trail was made into the Valley on the Mariposa side, and the first hotel was opened in 1857, when regular pleasure travel commenced. Many men eminent in the pursuit of science have made careful geological studies and examinations of the Valley, and have arrived at different theories regarding its formation. Some pretend to trace it to glacial disturbances; others claim that it is the result of erosion; while still others adopt the theory that it is the result of a vast rent or fissure. Major Truman says that none of these theories are well sustained, but that the most natural as well as the most popular explanation of the formation is that during some convulsion of nature, "or something else of that kind," its bottom fell out. All tourists, explorers and geologists agree that the scenery of the Yosemite is of a type peculiar and unique.

Boating on Donner Lake.
FROM NORDHOFF'S CALIFORNIA, PUBLISHED BY HARPER & BROS.

·The other attractions of California are its geysers, mountains, lakes, and big trees. Perhaps the Geysers, situated in Sonoma County, 100 miles from San Francisco, are, partly owing to their accessibility and partly on account of their fame as objects of wonder, more generally visited than any other Pacific Coast attractions, the Yosemite excepted. From the largest to the smallest, from the "Steamboat" to the "Witches' Caldron" down to infinitesimal bubbles to be seen in every direction, from the mouth of the seething, boiling, trembling cañon to its head there are at least a hundred springs, of all shapes, colors, conditions and temperatures. On every foot of ground alum, magnesia, tartaric acid, epsom salts, ammonia, nitre, iron and sulphur abound, being constantly sputtered out from caldrons of black, sulphurous, boiling water. At thousands of orifices you find hot, scalding steam escaping, and forming beautiful deposits of snowy sulphur crystals. The tourist can hardly form conclusions from a descrip-tion of this Plutonian realm, this branch of Hades, nestling among umbrageous oaks and firs, this prodigious laboratory and olla podrida of liquids and salts. With its "Devil's

Kitchen,'' its " Devil's Inkstand," its " Devil's Armchair," and its " Devil's Machine Shop,"
this " Devil's Cañon " is a devil of a place, and the injunction of " Don't you forget it " is
unnecessary. The fame of the big trees of Calaveras and Mariposa groves is known to every
schoolboy. These enormous giants of the forest grow so large that theatrical performances
may be given on their stumps, and stage-coaches driven through holes cut in their trunks
while still standing.

> " The giant trees, in silent majesty,
> Like pillars stand 'neath Heaven's mighty dome.
> 'Twould seem that, perched upon their topmost branch,
> With outstretched finger man might touch the stars."

Observation Car.
FROM NORDHOFF'S CALIFORNIA, PUBLISHED BY HARPER & BROS.

Hittell, in his " Resources of California," says : " One of the trees which is down—the
Father of the Forest—must have been four hundred and fifty feet high and forty feet in diameter. In 1853 one of the largest trees, ninety-two feet in circumference and over three
hundred feet high, was cut down. Five men worked twenty-five days in felling it, using large
augers. According to Mr. Hutchings's statement, the Calaveras Grove of Big Trees was the
first one discovered by white men, and the date was the spring of 1852. The person who
first stumbled on these vegetable monsters was Mr. A. T. Dowd, a hunter employed by the
Union Water Company to supply the men in their employ with fresh meat, while digging a
canal to bring the water down to Murphy's. According to the accounts, the discoverer found
that his story gained so little credence among the workmen that he was obliged to resort to a
ruse to get them to where the trees were."

Foremost among the lakes of California—of which there are many folded in the mountain-
tops like emeralds in their setting—and ranking all others in point of rare beauty and situation,
is Lake Tahoe. It is a magnificent sheet of water, twenty-five miles in length and in some

places from twelve to fourteen miles in width. It has a depth of 1700 hundred feet, an altitude of 6216 feet, and is surrounded by mountains which tower above the lake from 2000 to nearly 5000 feet. More might be said of Tahoe, perhaps, than of any other spot in California—excepting, always Yosemite. There are grandeur and enchantment at all times in the scenery which environs the lake, and the panorama of mountain and valley, meadow-land and woodland, sunshine and cloud, as viewed from Tahoe City, is spacious, inspiriting and impressive. The summer sunsets upon Tahoe are remarkable for their great beauty and wealth of coloring, and are pronounced by European tourists as superior to those so often mirrored in Lakes Como and Maggiore. Donner Lake perpetuates the name of George Donner, an early emigrant, who, with his wife and a large number of other men and women belonging to an expedition, were overtaken by a tremendous storm of snow early in the winter of 1846, during which many perished, at a point upon the old stage-road not far distant from this beautiful body of water. Some years ago a well-known California writer produced a volume entitled "Fate of the Donner Party," in which he apostrophizes this enchanting lake as follows : " Three miles from Truckee lies one of the fairest and most picturesque lakes in all the Sierra. Above and on either side are lofty mountains, with castellated granite crests, while below, at the mouth of the lake, a grassy, meadowy valley widens out and extends almost to Truckee. The body of water is three miles long, one mile and a half wide, and 483 feet in depth. Tourists and picnic parties annually flock to its shores, and Bierstadt has made it the subject of one of his finest, grandest paintings. In summer, its willowy thickets, its groves of tamarack and forests of pine are the favorite haunts and resting-places of the quail and grouse. Beautiful speckled mountain trout plentifully abound in its crystalline waters, which reflect as in a polished mirror the lofty overhanging mountains, with every stately pine, bounding rivulet, blossoming shrub, and waving fern."

The tourist who would see California at its best should visit it in the spring. With the month of June the dry season sets in, and vegetation becomes parched and dusty. In March, April and May the country is at its loveliest. But portions of the State are latterly much sought as winter resorts. Los Angeles, Ventura, Santa Barbara, San Diego, and San Bernardino counties, all in the southern part of the State, form what is generally known as "Tropical California," a land

> " Where a wind ever soft from the blue heaven blows,
> And the groves are of laurel and myrtle and rose."

Where luscious fruits of many species and unnumbered varieties load the trees, and gentle breezes come through the bowers. Much has been written of the influence of external nature upon national character. It is considered as established that extreme cold dulls the intellect ; that extreme heat debases morals and enervates the body ; that the temperate zone only can produce a really high and pure civilization. It has further been noted that the people of mountainous countries are, other things being equal, superior to the people of level countries, and the dwellers on the sea-coast to those of the interior. The Californian, like the Greek, has every advantage of natural surroundings. He is neither dulled by extreme cold nor demoralized by extreme heat ; he aspires with the mountains ; he drinks in the many sounding sea, figuratively speaking ; actually he has something better to drink in his clear waters and the juices of his luscious grape. In other parts of the temperate zone men get more than an occasional taste of both the torrid and the frigid ; in California it is not so. The Pacific slope enjoys warmer winters than the Eastern States, and cooler summers. The nights are always cool ; the days never oppressively sultry. There are no violent storms of any kind ; the air is dry and invigorating. To reach California take Pennsylvania Railroad to Chicago, whence two routes are offered: the Rock Island and Union and Central Pacific, or the new southern route via Atchison, Topeka and Santa Fé and Southern Pacific. Either is a long, tedious ride, though simple holiday amusement compared to the methods of crossing the plains a few years ago.

The Catskills.

It is but justice to say that there is, probably, not a mountainous region on the globe more picturesque and varied and more naturally the home of romance and tradition, than that of twenty or thirty miles square which embrace the rare Catskills. If one approach them from the Hudson, his first glimpse will show Round Top and High Peak towering into the sky, with the other mountains gathered about them, as children about their parents. The ascent by the stage-route, from the village of Catskill, is so easy as to seem, at first, tame; but the charm of the way soon disposes of any such sentiment. The fine, full-foliaged trees in Rip Van Winkle's dell make a pleasant period for the backward views; the noble North Mountain continually rises before and gives dignity to the scene; vistas of blue-browed hills stretch out before to an unexpected reach; the walk up-grade is beguiled by the music of invisible waterfalls, while the tender sigh of the woods and the sweet breath of the flowers linger in the sylvan cool, and a peaceful spell broods on the dreamy outlooks. Eight miles ahead one steps from behind the large hotel on the landing and from the platform looks down upon a view as original as superb. The climbing of the mountain has been so natural and the ascent so cunningly covered by the hand of nature, that it is bewildering and delightful to be thus suddenly perched 2700 feet above yon distant shaving-like Hudson, and look down into this royal sweep. On one side mountains and ravines, gorges and dells, glittering waterfalls and shining brooks, all framed in the deep green of the grand forests, with here and there a touch of color in a clump of flowers. On the other side, sheer down at the bottom of the precipice, and at the foot of the mountains, the apparently flat surface of the valley spreads out to the Hudson, which rolls out its slender, silver length for fifty miles; here toy houses and tiny buildings, the quiet homes of those toiling farmers, who look like insects crawling over the plain below. The two opposed sections as sharply separated as if an express order of the Maker had placed here the rugged and picturesque, there the fair and pastoral. Further out, the forests on their summits serve to mark the line

Inspiration Rock.

of hills stretching out towards Saugerties. Over the river the country seems to have been purposely placed on exhibition slope to show as much of the rolling surface as possible, while the horizon is circled by the Hudson Highlands, the Berkshire and the Green Mountains, which unite their chains in a line of blue that grows dim and distant in the gathering twilight.

At many points in the Catskills one may from his bed see the sun rise a hundred miles away, glorify the distant summits of the Green Mountains in Vermont, sparkle on the White Mountains, light up the rich Connecticut plains, and then flood the whole ten thousand square miles that lie within the range of the eye. From Table Rock, on North Mountain, may be had a magnificent view of this landscape from one standpoint. From South Mountain, one may see the Catskill Pass and the peaks of New Jersey, while the ambitious may climb so high as to imagine he sees, beyond the intervening beauties, the city of Albany,—an anthill in a

The Wittenberg, from Mount Cornell.

meadow. The mountains have pilgrimages innumerable, and exquisite nooks in abundance. In the region of Round Top and High Peak are the two lakes, North and South, whose common outlet falls into a deep cleft, the first fall one hundred and eighty feet, the second eighty feet, and the third forty feet. The falls are seen to advantage from below, where the walls behind rise, rugged and broken, three hundred feet. The supply of water being limited, a dam has been placed across the verge of the cliff, and ordinarily a thin ribbon drops over, but at certain periods the dam is opened, and the body of water dashes down in spirited style, the curling spray flies back into one's face, and out amongst the big boulders, half-hidden by nodding ferns, the red-capped rubus and tender-tinted laurel bushes, the Catterskill bounds and sparkles from the cool, dark depths, to wind its devious way eight miles to the Hudson, which it enters near the village. After an hour or so spent in exploring this fine glen, a delightful walk to the Clove, a mile distant, may make a charming period to the initiative excursion, or, if one choose, he may ramble a mile and a half to Sunset Rock, which commands some noble views.

One of the most romantic resorts in the Catskills is Haines's Falls. In the first leap of one hundred and fifty feet, and the second of eighty feet, the water is churned and broken up into a white, angry mass, which continues in its downward course a quarter of a mile, in which space the stream is lowered four hundred and seventy-five feet. The way down from the foot of these beautiful and varied Falls is through the Kaaterskill Clove, a ravine so rare as to form a fitting station between the laughing waters and the plain beneath. Here are the curved and tumbled High Rocks and the Fawn's Leap Falls. The edge of these Falls sweeps

Mount Cornell, from Wittenberg.

around in a fine curve, that seems like a heavy piece of masonry work, while the water pitches thirty feet into an immense pit of granite. At the mouth of the Clove lies Palensville, a rail-road terminus, and six miles from the town is Plattekill Clove, reached by a rough road. The principal feature of this Clove is the Black Chasm Falls, three hundred feet high. A ride on the railroad through Stony Clove, some six miles distant, gives a good idea of mountain engineering, and shows some interesting and wild perspectives. Four miles west of the entrance to this Clove, Hunter Mountain rears its head four thousand and eighty-two feet.

The most prominent of the Catskills is High Peak, six miles from the main hotels. The trip is generally undertaken by the venturesome, as the way is rough and hard, but the magnificent view from the summit (thirty-eight hundred and four feet high) of the combined out-

The Overlook from West Hurley.

looks well repays the toil. The southern portion of the Catskills is not so well known as the region more particularly alluded to above, but abounds in lofty spurs, such as the Storm King (four thousand feet high), Cornell Mountain, Overlook Mountain, etc., while spots, like the Poet's Glen, Overlook Rock, Lover's Retreat, and the Pilgrim's Pass, are as charming as their names are suggestive.

The casual mention of these points of interest can give no idea of the riches held in store. One walk along the Cauterskill from Fawn's Leap Falls to Haines's Falls will reveal such a succession of beauties that many a lovely picture will linger in one's memory for years to come. Dewy rock-grottos open into others beyond, and everywhere sweet moss-laid nooks, where fairies might hold carnival under the shade of the brightly dressed, immaculate iron-wood, the broad-crowned alders, and the swaying mountain willows. The air is laden with woody perfumes, and the smell of the junipers, the cedars, the spruces, and the balsams, that sweep away all taint of the far-off city, and infuse new vigor into the frame of the weary worker. It is the one added spell to the charms of the vistas beyond and the forest around, where the beautiful white birch coquettes with the dignified oak and smiles on the blushing maple. The banks and by-ways are pink and white with the

Haines's Falls.

bloom of the laurel, and the ground is spread with an artistic rug of white pipsissewa trumpets, pointed with blue-eyed grass, and relieved here and there by clusters of pink ear-drops, maidens'-hair fern, the purple fox-glove, and many another delicate spray ; and down through these glens goes the Cauterskill, in and out, now murmuring around a gentle curve, and now breaking into a thousand rills at the brink of a precipice, to meet in a merry volume further down the brook.

What coy glimpses of beauty open as the stream dashes on its course! Nearly at the bottom of the gorge is a spot where one may sit in the shadows of the rocks, beneath the falls, and dream out a day. The sketches that many nature-loving artists have taken from this point tell but a part of the story one may read in the way of the brook, now silvery and glancing, now rainbow-hued, and waving arms of wind-tossed spray; in the witchery of the trembling bowers of softly lit foliage; in the fresh colors of starry flowers painted on a background of green-fringed rock; in the music of the birds, mingling with the song of the waters, when, over all, is that indescribable benison that rests upon one in the midst of nature's own retreats:

> " Midst greens and shades the Cauterskill leaps,
>   From cliffs where the wood-flower clings;
> All summer he moistens his verdant steps,
>   With the sweet light spray of the mountain springs;
> And he shakes the woods on the mountain side,
>   When they drip with the rains of autumn tide."

The Catskills are easy of access. Boats run daily from New York, the one leaving from the foot of Vesey Street at 8.30 A.M. arriving at Catskill landing at 3.20 P.M., or the traveller may go *via* the Hudson Railroad to Catskill station. From the North the morning. boat from Albany reaches the landing at 11 A.M , or one may go by rail, as above. Omnibuses from the landing to the village. A railroad from Catskill to Palensville, where Kaaterskill Clove debouches on the Hudson River valley, saves stage-ride to those. wishing to go to Pine Orchard, Round Top, and adjacent points. Another route is by the new Stony Clove Railroad, which connects with the Delaware Railroad at Phœnicia and runs to Hunter on Schoharie Creek, a distance of twelve miles. A stage-route from Tannersville Junction on this road leads to principal points. The mountains may be entered from the south by railway from Rondout to West Hurley, and thence by stage. A fee of twenty-five cents is usually charged at points of interest. The hotels and boarding-houses are numerous, and range in prices from $1.50 to $4.50 per day, and $10 to $25 per week.

The Adirondacks.

THE most distinctively mountainous section east of the Rockies is that tract stretching from Mohawk River on the south to Canada on the north, with the historic and beautiful Lakes Champlain and George on the east, and the clear St. Lawrence at the northwest. The mountains, to the number of five hundred, have been placed upon a plateau itself two thousand feet above the level of the sea, in five ranges, which cross in parallel lines from northeast to southwest, and rise in tiers toward the west, the highest mountains, Seward, McIntyre, McMartin, Whiteface, Dix Peak, Colden, Santanoni, Snowy Mountain, and Pharaoh, all nearly 5000 feet high, and Mt. Marcy, 5337 feet high, being in the most western, Clinton or Adirondack range. The mountains are remarkable for their uniformity in the matter of height. There are loftier spurs in both the White Mountains of New Hampshire and in the Black Mountains of North Carolina, but the Adirondacks have a higher average than either of these. In the valley between these ranges and mountains lie a thousand lakes, that mirror on their polished bosoms the steep and densely wooded declivities, and the stony summits above. Everywhere are these bodies of water, spread over a reach twenty miles long, or nestling in a hollow, pent up within the bounds of a few rods; in a basin in the raised floor of this region, fifteen hundred feet above the sea, or, as Lake Perkins, up in the clouds, three times as high. The largest of the lakes are the Fulton Lakes, the Saranacs, Tupper, Long Lake, Colden, Henderson, Sanford, Eckford, Raquette, Forked, Newcomb and Pleasant. Down the by-ways from lake to lake a maze of brooks and rivers join the waters of the mountains, flowing through the valleys, as the Saranac and the Ausable, in marking lines of silver to separate the ranges, or as the Bóreas, the Cedar and the Hudson rivers, cracking through rock-ribbed courses to meet in a common band below, that winds away to the waiting sea. These latter views clearly define the southern continuation of the valleys traced above by the Ausable, while Raquette Lake, Long Lake, and the Fulton Lakes follow the depressions from the Saranac to the south-west. In Raquette Lake rises the river of the same name—the river of the mountains; born

amongst them and of them, the pride and beauty of the Adirondacks, it sweeps along a devious course for one hundred and twenty miles, and finally mingles its sweet waters with those of the St. Lawrence.

There are half a dozen different routes into the Adirondacks and the Wilderness, and one may follow his fancy or suit his convenience by travelling along the line of the mountains, or climbing across the ranges. At the northeast, the sentinel of the system is Whiteface Mountain, which looms up from the Wilmington side, superb and grand, a mountain view rarely equalled. From the summit of Whiteface one looks out to peaks beyond and mountains about, over forests charging up the heights, and far down the valleys to the south, beyond lovely Lake Placid, and down at the north side of the mountain on the jagged, deep and narrow chasm, the "Notch," through which the turbulent Ausable leaps in a series of rapids and cataracts. One might make a lengthy sojourn in this neighborhood, climbing the mountains to gaze upon the grand scene, looking at the Monarch himself from a boat on Lake Placid, visiting Paradox Pond,

Indian Pass.

whose outlet at high water flows back into the pond, or exploring Saranac Lake, a beautiful sheet, seven miles long, and having fifty-two romantic islands, wooded to the waters' edge, where the hemlocks wave their feathery arms in beckoning to the shadows at their feet. A very pleasant escape might be made by going down the Saranac River to Round Lake, a pretty, island-dotted circle of water, over which at times, the most terrific storms rage.   Tracing around the curve of the river one enters the Upper Saranac Lake, the largest of the Adirondack lakes, due west from Lake Placid and Saranac Lake, eight miles long, and from one to three miles wide.   Some few miles to the north, past the half-way Clear Lake, and beside the mountain of the same name, is St. Regis Lake, one of the most picturesque of the group, surrounded by numerous ponds, on the outskirts of civilization, and connected with Upper Saranac Lake by the "Route of the Nine Carries."

Grand Flume, Ausable Chasm.
FROM STODDARD'S GUIDE TO THE ADIRONDACKS.

To the southwest of the section described is a region filled with all that can interest the sportsman or delight the lover of nature. By steamer down the beautiful Raquette to Tupper Lakes, and one is fairly started on the round. Into this lake the Bog River, rich with speckled trout, drops in a charming cascade.   Up this river and beyond several "carries" is the lonely and sequestered sister of the former lake, Little Tupper Lake, whose gentle waves lap on a precipitous and rocky shore.   Then past a series of ponds and "carries" and the Raquette Falls to Long Lake, a watery seam in the valley for nearly twenty miles, from which one may see Mt. Seward rearing his head above the rolling plateau between. Farther south the lovely Forked Lake and the final Raquette Lake, the home of a host of wild birds and beasts. This whole section, including the southwestward Fulton Lake and the surrounding chain of eight lakes, is rich in varied scenery and mountain fastnesses, and abounds with game.

Opposite Port Kent on the Champlain and three miles distant is Ausable Chasm, a wild and beautiful cut through which the Ausable, after dashing seventy feet over the Birmingham Falls and then leaping the Horseshoe Falls, flows between walls a hundred feet high and fifty feet apart, in a channel at places but little more than two yards wide. The chasm is made easy of access by a stairway of a hundred and sixty-six steps, and throughout the nooks and rocks and pools are guarded by rails and fences as in the similar Watkins' Glen.   To enter the central eastern part of the Adirondack region one leaves West porton "Northwest Bay," passing Hurricane Peak, the Giant of the Valley, Bald Mount to the right, round-topped Cobble

Hill, and the Roaring Brook Falls, where a mountain stream dashes over a precipice five hundred feet high, to reach the monarch of them all—Mt. Marcy. A hard climb up the picturesque trail to the summit discloses the most magnificent view to be obtained amongst the mountains. The great peaks filing away in splendid ranges, and rising and falling in the distance, lakes studded with green-fringed islets and encircled by dense, heavy-foliaged forests, river and brooks chaining the lakes in rare bands, glancing in the sunlight and leaping over beetling cliffs, great gorges and wild chasms splitting through the flanks of the mountains or opening down into the bottom of the plateau. One must himself stand upon the lofty height and look out upon all these wilds, with the Green Mountains of Vermont and Lake Champlain in the fore and background, to picture the indescribably grand landscape.

Coming up to Mt. Marcy on the other side and from the south, the most notable body of water passed is Schroon Lake, a delightful resort in itself. On this side of Mt. Marcy are some of the most prominent mountains in the system, and many noble views. The trail up leads by Avalanche Lake, a very high and lovely sheet of water, and that unique and stupendous gorge, Indian Pass, in the most savage part of "Conyacragu," or Dismal Wilderness. This section is the wildest and most difficult in the Adirondacks, explored only by the adventurous sportsman, who at any step may have to look along the barrel of his rifle into the eyes of a black bear, a wolf, a panther, or a lynx. In the centre of this pass, 4000 feet above the level of the sea, rises the Hudson from the midst of rocky

Ausable Pond.
FROM STODDARD'S GUIDE TO THE ADIRONDACKS.

recesses, where winter lingers through the year, and close beside the source of the great river are the springs from whose cold depth the Ausable rises, so close beside, in fact, that "the wild-cat, lapping the waters of the one, may bathe his hind feet in the other; and a rock rolling from the precipice above could scatter spray from both in the same concussion."

In a brief sketch it would be impossible to enumerate the various points of interest and the many spots where the smoke has risen from camp-fires in the thirty years since the Adirondacks were explored. Suffice it to say that Nature has furnished the Wilderness in a manner to suit the most cosmopolitan taste. A party may find recreation and enjoyment in the neighborhood of the many good hotels, and pass the time in the orthodox pursuits common to mountain resorts; or a company of good spirits may don red flannel shirts and cowhide boots, and with guides row down the rivers and across the lakes, througe files of flags and grassy shallows, or shoot along the rushing rapids to float out into beds of the pickerel-flower, past banks lined with white and gold lilies, that load the air with perfume, and paddle at eventide toward some little bay over rustling rushes and spongy pads. Then begin the free, the joyous outdoor life. The axes ring out and the echoes wander through woods that mayhap never caught the sound before. The spruces, cedars and pines about contribute to the planting of the tent, and beneath the oak, linden, birch, poplar or fir the camp-fire throws out a ruddy glow. All about are the magnificent forests, with sturdy giants, jungles of undergrowth and prostrate monarchs that once sighed amongst their kind without a human ear to hearken. What a paradise for a sportsman! To row out on some lovely lake by golden sands and patches of lilies, with the fragrant breath of the balsam and the pine in the air, and have the guide send the boat cleaving into a narrow opening overhung by bushes, and there in the lily-lined and gold-flecked stream of black, slow-running water to see the sweet vista broken everywhere by leaping, splashing, splendid spotted trout. Ah! here and there they rush, cleaving the surface in hot pursuit of a dancing gnat, or jumping clear out of the brook to seize a passing fly. Then to come to rest in some steady pool, around a tufted bank and with the trees hanging out their branches overhead, uncoil the leader and cast out the flies, and as in a gleam of yellow light the hackle disappears strike down the pliant rod to fix the hook, and then play and humor and control the game fish as he whirls in his mad course around the pool. The fine lance-wood curves and quivers, and the silken hair whirls over the reel, but skilful management brings him at last to the surface gasping, to be scooped up in the landing-net and breathe his last at the bottom of the boat. To those who prefer the rifle to the rod the forests offer many attractions, and paramount to all, deer hunting. One may steal out at night and with muffled oar paddle noiselessly along the borders of a lake, till a dark outline ahead indicates a deer. As the lantern is opened the bright ray shoots across the waters and the animal looks up in momentary bewilderment. No one who has not at such a time held in his hand a breech-loader, and at the click of the trigger-seen the deer bound away in the line of the gleaming sights, can appreciate the thrill that courses through him as the sharp report rings out on the night air, nor the exultation that rises into a cheer, if the report is followed by the crash of the falling deer. All the caution and cunning and skill of an experienced hunter, however, are needed to often enjoy the pleasure. But without the hunting and fishing there are many ways of spending the days—exploring the nooks and corners of the lakes and ponds, running the rapids of some dashing streams, or admiring the grand scenery as it opens before the boatman. To know the delights of a savage life one must leave civilization behind, and in the heart of the wilderness drink in health and strength and be glad in perfect peace and forgetfulness.

The White Mountains.

WHILE the Alps, the Sierra Nevada, the Yosemite, and other mountain regions may contain higher mountains, deeper valleys, broader lakes, more extensive vistas, yet there is nothing to rival the White Mountains in their infinite variety of scenery, manifold kaleidoscopic combinations of natural grandeur, and landscape effects; the contrasts and brilliance of color, too, varying not only with the seasons, but with the changing hours of the day. Their valleys and glens, redounding with historic interest, unlike the unoccupied forests beneath the peaks of the Rockies or the desolate glaciers of the Alps, have been the sites of towns known to many generations, and are still occupied by the hardy descendants of the ancient conquerors of both wilderness and a savage foe. The comparatively ready accessibility of this truly wonderful region, with its inexhaustible supply of rich material for every tourist, whether he crave sensational effects, high artistic pleasure, wild rambles, or grand solitude—with its stupendous mountains, hanging rocks and crags, crystal streams, verdant woods and meadows, grand cascades and roaring torrents, deep ravines, and beautiful valleys and lakes—renders it an inexplicable surprise that so many American people should cross the ocean to admire scenery most of which is inferior to this charming portion of New England.

Mount Washington, from Fayfan's
FROM "THE HEART OF THE WHITE MOUNTAINS," PUBLISHED BY HARPER & BROS

It will be the endeavor of this chapter to hastily conduct the tourist through every way in this grand and picturesque region, and point out the principal attractions and places of interest. It may be noted at the outset that excellent hotels and boarding-houses will be found in every village and hamlet ; and at no place will the visitor find the country lacking in this respect.

Occupying the northern portion of New Hampshire, and within a half day's ride from Boston, are these highlands called the White Mountains, comprising two clusters or groups of peaks, locally known as the White and Franconia Mountains, divided by table-land from ten to twenty miles wide. A ride from Boston *via* the Eastern Railroad to Conway, one hundred and thirty-two miles, will bring the tourist to this beautifully situated village, which serves as an excellent centre for many short and interesting excursions. To the visitor preferring the

Crystal Cascade

FROM "THE HEART OF THE WHITE MOUNTAINS," PUBLISHED BY HARPER & BROS.

air of rural quiet to the social attractions and brilliant life of its fashionable neighbor, North Conway, it affords great advantages. Five miles to the north is the summer capital of the mountain region, North Conway, one of the prettiest towns in New England. It is a favorite rendezvous for artists and the fashionable world, and very largely frequented throughout the best part of the season, which the Rev. Thomas Starr King says is "from the middle of June to the middle of July." The very entrance into North Conway seems like the introduction through a beautiful gateway of mountains into the retreats of nature—grand, imposing, entrancing. Admirable views present themselves on all sides. Looking up the village street Thorn Mountain is seen, behind which lies Jackson, and farther on, up the Ellis River valley, Gorham and Androscoggin. To the right the gentle slopes of the Kearsarge rise, with the silver-gray crest of the mountain towering to the clouds; to the left the Ledges and Moat Mountain present themselves, the abrupt declivities of the latter forming a fitting termination to the picturesque scenery of the beautiful valley beneath it. Following the old stage-road from North Conway in a northwesterly direction the tourist finds beautiful prospects all along the route as he passes through the Cathedral Woods, past the Intervale House, with Mount Kearsarge to his right,

Moat Mountain on the left, and with the most charming views opening into the Saco intervales. Soon Thorn Mountain is passed on the right, the Ellis River, the runaway from Mount Washington, is crossed, and the town of Jackson is reached. This hamlet is very prettily situated and a favorite resort, affording fine views of Tin, Thorn, Moat and Iron Mountains. The Jackson Falls, within the village, and visible from the highway bridge over Wild Cat Brook, present a pretty scene, the water being precipitated in glistening white bands over a high dark ledge into foaming pools below. Trout-fishing in the brook is one of the favorite pastimes of the many tourists sojourning here. Beautiful views of Mount Washington are obtained from the Fernald and Prospect farms, near Jackson. Between Jackson and Goodrich Falls, one and a half miles below, the prospect is one of the finest in the highlands. The Carter Notch will repay the tourist well for the time spent upon visiting it. From here the visitor may take the stage for the Glen House. This hotel is at the very base of the monarch of the White Mountains, Mount Washington. Luxuriant forest scenery opens on every side as the traveller progresses into the Glen, which is three hours by stage from Glen Station. The latter point can be reached *via* the Grand Trunk and Eastern Railroad, or by any of the routes over Mount Washington. The five highest peaks of the White Mountains (Madison, Adams, Jefferson, Clay and Washington) present themselves in grand array from this point. The disciple of Walton may enjoy good trout-fishing in the vicinity, while the general tourist will find the neighborhood abounding with points of interest, chief among which are the Garnet Pools, the Imp, Thompson's Falls, the Emerald Pool, the Glen-Ellis Falls, the Crystal Cascade, and Carter Dome. From the Glen House the road leads along the Peabody Valley, a distance of eight miles, to Gorham, the nearest village to the great peaks north of Mount Washington. "No point in the mountains," says Thomas Starr King, who spent several seasons in Gorham, "offers views to be gained by walks of a mile or two, that are more noble and memorable." The village, which is an important station of the Grand Trunk Railroad, is 812 feet above the sea, and is located in a broad valley, whose dry, bracing air is healthful and invigorating. East of Gorham, near the railroad station (Grand Trunk line) of Shelburne, is spread, over a rugged and mountainous area, the little hamlet of Shelburne, the road through which is, according to Mr. King, unsurpassed by any drive of equal length among the mountains, for varied interest in beauty of scenery, historic and traditional associations connected with the prominent points of the landscape, and the scientific attractions of some portions of it. The chief mountains of the town are the Ingalls, Baldcap, and the northern peaks of Moriah. Mount Winthrop, within the town, affords an excellent point for overlooking the Androscoggin Valley, with the Newry range and other mountain heights of Western Maine in the distance. Continuing the trip in the opposite direction from Gorham, the nearest station reached is Berlin Falls. The Androscoggin River descends here, nearly 200 feet in one mile of its course, in powerful falls and rapids, the most interesting being the Berlin Falls. Other noteworthy points are the Alpine Cascades and Mount Forest. A favorite drive with visitors to this locality is that over the Milan road along the river through its picturesque valley to Milan, eight miles distant. From West Milan and Green's Ledge grand views of the mountain ranges and the Androscoggin Valley reward the tourist for the trip.

Returning from this northwardly invasion into this beautiful region of the White Mountains towards its grand centre of observation, North Conway, the Portland and Ogdensburg Railroad may serve as a basis of further operations, with the little hamlet of Upper Bartlett for a starting-point. This place is admirably situated, being entirely surrounded by mountains. There are the Carrigan, Nancy Range, Tremont, and Haystack, on the west ; Hart's and Willoughby Ledges, Parker, Crawford, Resolution, Langdon, and Pickering, on the north ;

Kearsarge and Moat to the east ; and Table and Bear Mountains to the south.   Fine trouting
in the tributaries of the Saco tempts the angler, and from here the fine view of Mount Carrigan is had which Champney's famous painting has made so widely known.   Proceeding
further, Bemis Station is reached, another centre of rich mountain scenery.   Near the station
is the old Mount Crawford House, whose site was occupied before the year 1800 by Abel
Crawford.   It was formerly one of the chief hostelries of the mountain region, but has long
since been closed to the public.   Interesting excursions may be made from here to the Crawford and Nancy Mountains, the latter of which derives its name from a romantic though sad
incident in the early history of the neighborhood : In the winter of 1788, Nancy, a servant in
the family of Colonel Whipple, was about to go to Portsmouth with her lover to be married.
In her simple trustfulness, the young girl confided the small sum, which constituted all her
marriage portion, to her betrothed, who repaid her with the basest treachery.   Seizing an
opportunity, her faithless admirer left the hamlet.   On learning this, she set out after him
through the snow, reaching the camping-place at the Notch, thirty miles distant, only to find it
deserted, and herself unable to rekindle the smouldering fire.   She pressed on until, after crossing and recrossing the icy Saco several times, she died of utter exhaustion upon the south bank of

Owl's Head Mountain.

Nancy's Brook, where, under a canopy of evergreen which the snow tenderly drooped over her,
she was found by men who had gone in search of her.   The lover became insane and died, a few
years afterward, a raving maniac.   The view from the Crawford House is particularly grand,
with the pleasant Crawford Glen in the foreground, and many of the loftiest peaks of the region
beyond.   Between Willard, Willey, Webster, and Jackson Mountains (all of which may be
seen from here), and dividing the great New Hampshire group of mountains near its centre,
is a deep pass, the White Mountain Notch.   The massive walls are seen towering to a height
of two thousand feet, and, indeed, some of the highest peaks are lost to sight among the clouds.
The base of the Notch forms the bed of the wild, impetuous Saco River, which descends
through rocky débris of old avalanches, and winds about and dashes and splashes over huge
boulders along this vast ravine.   The splendor of autumnal scenes in the White Mountain
Notch has been again and again enthusiastically described by the pen of the author and portrayed by the brush of the artist.   A number of falls, notably the Flume and Silver Cascades,
and the Ripley and Arethusa Falls, charm the visitor, and invite him to prolong his stay here.
The Crawford House occupies the supposed bed of an ancient lake, upon a plateau nineteen

hundred feet above the sea. Near by are also the Gate of the Notch, the Elephant's Head, Bucher's Cascades, and Gibbs's Falls. Visits to Mounts Willard and Willey, from which beautiful views are obtained, are among the most pleasant and profitable tours to be made in this region. Four miles north of the Crawford, is the Fabyan House, fifteen hundred and seventy feet above the sea. Most of the summits of the Presidéntial range of mountains are visible from here. The Ammonoosuc Falls, where the stream descends over rapids for some distance above and then makes a fall of nearly fifty feet through a narrow gorge, whose walls are polished ledges of granite, the Giant's Grave, a mound of river gravel or a sandbar formed by the reaction of the ocean-waves against the adjacent hills, and many other points invite the tourist's attention. Good trout-fishing forms additional attraction. The Twin Mountain House is located upon a terrace of the Ammonoosuc River, about five miles west of the Fabyan. The Twin Mountains, which are difficult of access, are best seen from Mounts Washington and Lafayette. The best point of advance is considered to be the head of Little River. Eight miles west of the Twin Mountain House is Bethlehem Station. The usual approach to this point from the south is *via* the Boston, Concord and Montreal Road, and its Mount Washington branch. Passengers from North Conway go by the Ogdensburg Railroad through the Notch. Stages from the hotels will be found in waiting at the station. The village of Bethlehem Street, on a high plateau, fourteen hundred and fifty feet above the level of the sea, is said to be the highest village east of the Rocky Mountains. The view from here is broad and imposing, and the surroundings lack little if any of the beauty of those of North Conway. The drives in the neighborhood are varied and delightful. Bethlehem is particularly sought by victims of hay fever, to whom it is a perfect harbor of refuge, while its sanitary advantages in other respects have made a sojourn here a frequent object of recommendation to other invalids. The Mount Washington House, delightfully situated a few rods from the main street, is one of the best caravansaries in the highlands, and is commended to sojourners at Bethlehem. Eight miles north of Bethlehem, on the St. John's River, is the pretty town of Whitefield. The Howland Observatory is two miles distant from the village, and commands a grand view. Dalton is next reached at the head of the Fifteen-mile Falls of the Connecticut River, a chain of wild rapids in a narrow valley. Farther on, towards the northwestern verge of the mountain region, one of the most beautiful villages of the White Mountains, Lancaster, is located. It has a delightful climate, and is surrounded by some of the best farms in the State. It lies on the Israel's River, near its confluence with the Connecticut. Of its beautiful valley, Sir Charles Dilke says: "The world can show few scenes more winning than Israel's River Valley." Pleasant drives in the vicinity of the village offer many advantages for extended excursions, and afford magnificent views, extending over the rich meadows and fruitful fields, and along the rivers to the distant mountain background.

Turning back into the very heart of the northwestern region of the White Mountains, the hamlet of Jefferson Hill is encountered, eight miles southeast of Lancaster. Thomas Starr King says of it, enthusiastically: "Jefferson Hill may, without exaggeration, be called the *ultima thule* of grandeur in an artist's pilgrimage among the New Hampshire mountains, for at no other point can he see the White Hills themselves in such array and force." From here many tours into the surrounding country, with grand views of the Presidential range of mountains, can be made. Visits to Mount Starr King, and Owl's Head, on Cherry Mountain, are especially interesting, as is also the ascent of Mount Adams and the drive to the top of Randolph's Hill. The invasion into the ranks of the gigantic cluster of beautiful mountains, the Presidential Range, is thus made. A thousand wonders of the mountain-world lie open to the tourist in his rambles among these lofty peaks. The Lakes of the Clouds, five thousand feet above the sea, between Mounts Washington and Monroe ; the Falls of the Ammonoosuc River, which rises here, and descends over two thousand feet in the first three miles of its course; that stupendous declivity, Tuckerman's Ravine, with its wonderful snow-arch, formed by the solidification of the ponderous masses of snow driven into it by the winter storms, and filling it, while the mountain streams, passing beneath the icy cupola, carry off the snowy contents beneath, leaving the ice-arch supported by the huge boulders of the ravine; the castellated ridge of Mount Jefferson, that invincible fortress of nature's own construction; the Great Gulf, that terrible yet fascinating gorge, with its wide-split crevasses, from which the encircling mountains appear to ascend, as if out of the mighty depths of the earth itself; Hermit Lake, King's Ravine, the Alpine Garden,—all these, and many other, points of interest awe and inspire with wonder, and delight the visitor. Mount Washington, the giant of the mountain range, is the highest point on the North Atlantic coast, its lofty peak rising to an altitude of sixty-two hundred and ninety feet above the level of the sea. Its summit has the arctic climate

Echo Lake.
FROM "THE HEART OF THE WHITE MOUNTAINS," PUBLISHED BY HARPER & BROS.

of Greenland, though twenty-six degrees farther south than the latter, and is above the limit of the tree-growing region. Pages of interesting description might be written, and have been written, of the wonders seen from this elevation, penetrating, as it were, into the very secrets of the heavens. Thunder and lightning playing among heavy storms; rainbows, remaining for hours, with two and three supplementary bows; antherias and coronas, such as are never seen elsewhere in our latitudes; sunrises and sunsets of magic splendor; conflicts of the winds and clouds, the latter of wonderful varieties of shapes, colors, and movements; and frostwork of the most exquisite master character, are among the phenomena that may be witnessed upon this mountain, the spectator realizing almost the royal treasures with which wild fancy, in fairy stories, fills the interior of enchanted mountains, in the surroundings of this wonderful peak. The first ascent of Mount Washington was made by Darby Field, an Irishman, in June, 1642. Accompanied by two Indians, he started from Piscat (Portsmouth), accomplishing the journey in eighteen days. The view from the summit of Mount Washington eucompasses nearly one thousand miles, embracing parts of five States and the Province of Quebec. and sweeping over scores of villages and towns, hundreds of hills and mountains, and lakes, rivers, and valleys of New England. Provided with an abundance of heavy clothing, suitable for the arctic atmosphere of the top, the tourist may ascend the mountain by the convenient railway to the summit.

Five miles west of Bethlehem, at the busy little village of Littleton, the traveller will find himself again at a convenient starting-point for an exploring tour, this time directed toward the other division of the grand New Hampshire mountain range, the Franconia Mountains, which, while less imposing and majestic than the White, possess features equally wonderful, and often of greater beauty. The Franconia Notch, a valley five to six miles long, through which the Pemigewasset River pours its pure mountain waters, about ten miles south of Bethlehem, is reached by rail from the latter place, or by stage from Plymouth. The Notch is called by Rev. Starr King "a huge museum of curiosities." Harriet Martineau declared it the noblest mountain pass that she saw in the United States. Nearly a mile north of the Profile House, situated near the north end of the Notch, is the beautiful Echo Lake. From a boat upon this pretty sheet of water, the human voice is echoed and re-echoed with peculiar distinctness, and the report of a cannon or gun is answered by artillery and musketry from every mountain and hill. The singing of a song, or the playing of a tune upon a musical instrument, seems to make vocal the forests of every hill, or fill them with bands of musicians, reminding the listener, as the echoing notes die in the distance, of the lines of theEnglish poet laureate :

> "Oh hark! oh hear! How thin and clear,
>   And thinner, clearer, farther going ;
> Oh sweet and far, from cliff and scaur,
>   The horns of elf-land faintly blowing!
> Blow! Let us hear the purple glens replying ;
> Blow, bugle! Answer, echoes, dying, dying !"

One of the greatest curiosities of nature to be seen here is the Profile, or Old Man of the Mountains, a remarkable natural rock-sculpture of a human head, measuring some forty feet from the tip of the chin to the flattened crown, and bearing a perfect resemblance to an antique profile. It is formed by the ledges on the upper cliffs of Mount Cannon. The latter, the long and massive ridge forming the western wall of the Notch, derives its name from one of the ledges, which is so balanced that it presents the appearance of a cannon protruding from the parapet of some fortification. Profile and Moran Lakes, with Eagle Cliff, form additional points of interest to the tourist, and are easily accessible. From here, also, the ascent of Mount Lafayette may be made over the so-called bridle-path. At the south end of the Notch is the Flume House. The chief point of interest in the vicinity is the Flume, a remarkable rock-gallery, some seven hundred feet in length, through which rushes a beautiful icy brook. Near the upper part of the Flume, firmly lodged between the narrowing walls, is a huge boulder. The Pool, the Basin, Mount Pemigewasset, and Georgianna Falls, in the neighborhood, are places of great interest to the sojourner here. Emerging from the Franconia Notch, and proceeding in his tour towards the southwestern portion of the mountain region, the visitor has reached the western part of the hill country, comprising what is termed the "Pemigewasset region," and completed his tour of the White Mountains country proper, which, while being marvellously grand, picturesque, and enchanting, in almost every feature, does not, as a writer has said, "by any means monopolize the beautiful landscape visions scattered through the New England States. Mount Washington is not the only peak worth climbing, nor are the Conway Meadows the only dreamland." Some of the other beauties and delights of New England to the tourist are treated under other heads in this work.

Mineral Springs Resorts.

" *Imprimis*, my darling, they drink
　　The waters so sparkling and clear ;
　Though the flavor is none of the best,
　　And the odor exceedingly queer ;
　But the fluid is mingled, you know,
　　With wholesome *medicinal things*,
　So they drink, and they drink, and they drink,
　　And that's what they do at the Springs.

　　　*　　*　　*　　*　　*　　*

" In short—as it goes in the world—
　　They eat, and they drink, and they sleep ;
　They talk, and they walk, and they woo ;
　　They sigh, and they laugh, and they weep ;
　They read, and they ride, and they dance,
　　(With other unspeakable things) ;
　They pray, and they play, and they pay,
　　And that's what they do at the Springs."

N various localities from Maine to California the surface of our country is dotted with mineral springs, around which throngs of people gather every season, some to drink the waters, others to mingle in the whirl of pleasure and society they find there. The question whether mineral waters are really beneficial medicinally has been extensively discussed, and in spite of the incredulity and ridicule so long indulged by a portion of the medical profession a verdict in favor of the waters has come to be generally accepted. Medical gentlemen now almost universally admit that patients afflicted with inveterate chronic diseases often resort to mineral springs with the result of a perfect cure. The pages of ancient authors frequently contain records of resorts where the sick bathed in healing waters or drank of medicinal fountains. In Greece the temples of Æsculapius were frequently erected near springs reputed to possess curative power. The ancient Athenians, during the summer months, sought the thermal-saline-sulphur baths of Ædepsus, in the island of Eubœa, about sixty miles by sea from Athens. The works of Latin writers contain frequent allusions to medicinal springs, testifying the esteem in which they were held by the Romans. In the brilliant days of imperial Rome, bathing formed a chief enjoyment of patrician and plebeian. The luxury of warm bathing was indulged in to such excess that at one time eight hundred thermal baths could be counted within the city, and several of these would accommodate three thousand bathers. Many of these structures covered entire squares, and were adorned with every architectural beauty. An approach to them showed beautiful marble porticos, supported by many-fluted columns, and within was a labyrinth of marble halls and colonnades, decorated with statuary and mosaics by the masters. Within the inclosure were gardens of rare flowers and exotics, and apartments containing well-arranged libraries and works of art. Americans do not yet treat their mineral baths upon this scale of magnificence, but they may be said to entertain almost an equal regard for the springs.

Saratoga.

As a fashionable resort Saratoga takes the lead of all others on this side of the Atlantic. During " the season " its mammoth hotels and numerous boarding-houses entertain an aggregation of humanity amounting to tens of thousands. The attraction which draws together this vast stream is chiefly the gayety of the place and its reputation among the votaries of fashion as the "swell" resort. The natural attractions are the group of mineral springs, the magnificent elms which shade the streets of the town, and the beautiful lake about four miles distant. There is also a race-course, among the finest in the country, and the summer races at Saratoga are noted turf events of the season. There is, withal, no more brilliant scene to be found in America than that presented at Saratoga in August, when the town is thronged with visitors, and thousands of private and public carriages join in the parade of wealth and style on Broadway and on the Boulevard. Broadway, the main street, extends for several miles, with the principal hotels near its centre, and a succession of costly villas beyond. The drives and promenades in the vicinity are delightful. The lake is nine miles long by three in width, and is a source of much pleasant amusement. It is reached by the Boulevard, which passes near the race-course and trout-ponds. In the near vicinity is a sequestered pond among the hills called "Lake Lovely." In the society at Saratoga it may appear that the gay and frivolous predominate, but it must be remembered that froth and foam come to the surface, while the still water rests quietly in its conscious power. The butterflies may sport in the sunshine,—we all love to see them, bright, golden-winged beauties as they are, glorifying the commonplace with their presence.

" Saratoga Society,
What endless variety!
What pinks of propriety!
What gems of sobriety!
What garrulous old folks,
What shy folks and bold folks,
And warm folks and cold folks!
Such curious dressing,
And tender caressing,
(Of course that is guessing,)
Such sharp Yankee Doodles,
And dandified noodles,

And other pet poodles!
Such very loud patterns,
(Worn often by slatterns!)
Such straight necks, and bow necks,
Such dark necks and snow necks,
And high necks and low necks!
With this sort and that sort,
The lean sort and fat sort,
The bright and the flat sort—
Saratoga is crammed full,
And *rammed* full, and *jammed* full."

Saratoga is situated thirty-eight miles north of Albany, 182 miles from New York, and 238 from Boston. It is reached from New York via boats up the Hudson to Albany and thence by the Rensselaer and Saratoga Railroad, or by the Hudson River Railroad to Albany. From Boston take the New York and New England Railroad to New York, or the Boston and Albany to the latter place. The springs are said to have been visited by invalids as early as 1773, but the principal spring was not discovered until 1792. It is said that the medicinal properties of the High Rock spring were known to the Iroquois Indians in 1535, and that Sir William Johnson was carried thither on a litter by the Mohawks in 1767, and he is believed to have been the first white man to visit the spring. The springs rise in a stratum of Potsdam sandstone, near a great break or fissure in the strata underlying the Saratoga valley, and reach the surface through a bed of blue clay. The waters are found very beneficial in affections of the liver, in some cases of chronic dyspepsia, and chronic diseases of the bowels. Besides other qualities, they appear to possess the virtues of a tonic united with those of a gentle cathartic. Most of the springs are now owned by stock companies, one of which has a stock capital of $1,000,000. Great quantities of the water are bottled and exported, and there is scarcely a town of any size in America in which they are not regularly sold. The process of boring artesian wells has been successfully introduced, and some of the most valuable of the new sources of water supply have recently been discovered in this way. The battle of Saratoga was fought here between the British, under General Burgoyne, and the Americans, under General Gates, commencing on the 7th of October, 1777, and terminating on the 16th, by the surrender of the entire

British force, numbering five thousand seven hundred and ninety-one men, with forty-two cannon and all their stores. The prisoners thus taken were held until the close of the war— more than five years. The present hotel system of Saratoga is unrivalled elsewhere in the world.

White Sulphur Springs, W. Va.

The Greenbrier White Sulphur Springs are, next to Saratoga, the best known and most popular of all the mineral springs resorts in this country. For many years they have been the resort par excellence of the South, and much sought by a select class from all sections. They are situated on Howard's Creek, in Greenbrier County, directly on the line of the Chesapeake and Ohio Railway, at the edge of the Great Western Valley and near the base of the Alleghany range of mountains, which rise at all points in picturesque beauty. Kate's Mountain, which recalls some heroic exploits of an Indian maiden of long ago, is one fine point of the scene southward ; while the Greenbrier Hills lie two miles away, toward the west, and the lofty Alleghanies tower up majestically on the north and east. Within the beautiful valley overlooked by these mountain summits is the magnificent hotel. In front the lawn spreads out, occupying probably 50 acres, and intersected by numerous winding walks. Encompassing the lawn on either side are long lines of shining white cottages, embowered beneath the shade of ancient oaks, while at the distant extremity, the famous spring bubbles beneath a pavilion. Taking one of the by-paths to the right, "Lover's Maze" is soon reached, and here, under a dense shade of forest trees, obscurely winding paths lead in every direction amid a thick growth of laurel, while precipitous declivities sink away, from which extended views of the deep valley below may be had, with the mountain ranges in the distance. Over all these natural beauties the "season" throws its spell of animation and revelry, for the White Sulphur is a place of much gayety, and pleasure-seeking reigns supreme.

These springs, according to a late medical writer, very much resemble the celebrated cold sulphur waters of Neuendorf in Electoral Hesse. They are beneficial in a wide range of diseases. It is not known precisely at what period the spring was discovered. Though the Indians undoubtedly knew its virtues, there is no record of its being used by the whites until 1778. Log-cabins were first erected on the spot in 1784–'86, and the place began to assume something of its present aspect about 1820. Since then it has been yearly improved, until it is capable of pleasantly housing some 2500 guests. The spring bubbles up from the earth in the lowest part of the valley, and is covered by a pavilion, formed of 12 Ionic columns, supporting a dome, crowned by a statue of Hygeia. The spring is at an elevation of 2000 feet above tide-water. Its temperature is 62° Fahr., and is uniform through all seasons. Its average yield is about 30 gallons per minute, and the supply is neither diminished in dry weather nor increased by the longest rains. The principal ingredients of the water are nitrogen gas, oxygen gas, carbonic acid, hydro-sulphuric acid, sulphate of lime and magnesia, and carbonate of lime. Its effect is alterative and stimulant, and it is beneficial in cases of dyspepsia, liverdisease, nervous diseases, cutaneous diseases, rheumatism and gout. To reach the White Sulphur from Washington take the Virginia Midland Railway to Charlottesville. From cities east and north take the Pennsylvania Railroad to Washington or steamers to Richmond. The distance from the latter point is 227 miles, and from old Point Comfort 321 miles.

Hot Springs, Arkansas.

IN a narrow valley, a mile and a half long, running north and south between the Ozark Mountains, in Garland County, Arkansas, lies the town of Hot Springs. It has an elevation of 1500 feet above the sea, and the surrounding region is wild and picturesque, while the town and immediate neighborhood are lovely with verdure during the greater part of the year.

The Hot Springs Creek flows past it, and the Washita River is six miles distant. There are 66 springs, which issue from the western slope of the Hot Springs Mountain, on the east side of the valley. 350 gallons a minute, or 500,000 gallons per day, at a temperature varying from 93° to 160° F., pour out into the creek, whose waters are sufficiently warm for bathing in mid-winter even. The springs are pure, so clear as to be transparent, are almost tasteless, and do not deposit any sediment. The water can be taken internally for their aperient and tonic effect, being highly recommended in blood diseases. The baths are beneficial in diseases of the skin, rheumatic complaints, and mercurial diseases, and are of three kinds : the *vapor baths* at 112° ; the *douche*, a spirit bath, at 120° ; and the *saving bath*, at 116°. The town and valley present a curious appearance to the incoming visitor, the rising steam from the springs giving the whole place the air of a great drying factory. The sun shines with intense ardor, but there is generally a draft through the valley which prevents the heat from becoming oppressive. The Hot Springs have of late been popular with many statesmen and other prominent men worn out with the exacting cares of a busy life. *Consumptives, however, and persons troubled with pulmonary or throat diseases should by all means avoid this section.* The Hot Springs are reached by the St. Louis, Iron Mountain and Southern Railroad, being 45 miles southwest of Little Rock. There will be but little trouble in securing quarters, as the entire town of 5000 inhabitants is in the business.

The Great Spirit Spring, Kansas.

ONE of the newer candidates for popular favor among the valuable mineral springs with which our country is dotted is the "Great Spirit," located near Cawker City, Kansas, on the Central Branch of the Union Pacific Railway. Though never much advertised or used as a moneyed attraction, the great medicinal value of this spring has been known to thousands of people for many years, and was famous among the aborigines long before the pale-face ever saw the beautiful Solomon Valley, wherein it is located. With the medical profession its waters have also had a high standing for several years. The formation in which the spring has its setting rises up like a vast cone, dripping the waters from an elevation of forty feet above its surroundings. According to the traditions of the Dakotas there were tribal ceremonies upon this spot in the ages of the past, when the great trees of Mariposa Grove were yet young, and it is said that the medicine-men of these tribes sent their patients long distances to cure their ills with this water. Arrows were sent from beyond the great pipe-stone quarries of Hiawatha to be consecrated in it. The locality has become a popular resort, both on account of the poetry and antiquity of the legends which cluster around it, and the reports of scientists, who have been led to analyze the water after reading the Jesuit records. It is still believed among the Indians that within the curious shrine of this spring the spirit of Waconda has remained since the battle of the seven suns, as the guardian for all time of the destinies of the brave fallen races of the red man, and that once in a hundred years, or thereabout, she will disturb the water and pour forth a flood to inundate and destroy the wicked.

Professor B. F. Mudge, assistant to Professor Marsh, of Yale College, visited the spring to study its geological place amongst the curious deposits of a region of limestone otherwise quite uniform. Prof. Mudge agrees with Dr. Adams, who also examined the locality, that the formation is tufa, quite calcareous, and that a greater amount of silica than most analyses present rises with the water ; so that, though the major portion of the rock is limestone, the flint (silica) is sufficient to make the mass of deposit intensely hard, and in some parts almost exclusively flint, the softer ingredients having apparently been worn away. Indeed, it is likely that the different relative quantities of lime and silica which the water has thrown up in different epochs of the history of the crust of the earth, as well as the varying quantities of mineral that would aid or impede the solidification of the lime, must have at one time built up the walls of the cone much higher than the present elevation of forty feet, and at other times reduced it to a level with the surface of the earth, or perhaps below the bed of the Solomon River, which is not more than three hundred feet distant, and all along which, within half a mile, there are huge blocks of this strange rock, that must have originated from the fissure out of which the water gushes, and whose bottom no pole has yet reached. But the cone at the spring itself is almost perfectly round, and its sides are not so steep as to prevent carriages from making the circuit near the water's edge. Formerly the water was found flowing about evenly over every part of the vast basin. This is the condition in which the Indians left it. The law of even formation seems to have been that wherever the thin sheet of water most frequently appeared there would be the most considerable deposit of the lime and flint, and a consequent elevation, while in no part would the overflow be great. Thus the chief escape would constantly vary, and give every point in the circle an opportunity to harden. There is a foliation of the solidified rock, and the people now living in the vicinity say it can be split with the stone-mason's axe.

There is no spring in this country having waters similar in character to the Waconda, or Great Spirit, except the High Rock at Saratoga, which partially resembles it. The sulphates of sodium and magnesia constitute twenty-five per cent. of the solids held in solution by this remarkable water, now so widely recommended by physicians. At San Filippo in Italy there is a spring of this character, which has built itself a protecting wall of rock more than two hundred feet thick. An analysis of the water of the Great Spirit Spring, by Professor Patrick, of the State University of Kansas, discloses the predominance of ingredients beneficial in all cases of liver and kidney difficulties. According to recent tests, it is believed among physicians that this water is the best known remedy for kidney diseases. This, coupled with the fact that the dry air of the plains has been found a potent remedial agent in pulmonary and malarial diseases, has directed special attention to the locality of this spring, and last season hundreds of people were there for whom accommodations could not be furnished. Many camped out for days in the neighborhood, and others travelled several miles daily to and from the nearest hotels. The owners of the property have never made any effort to utilize the spring further than to provide bath-houses and send away the water in barrels to those writing for it. But the unlooked-for demand for the water and for accommodations at the spring has led to the organization of a company, which will this season erect hotels and cottages and provide additional bath-houses. Those acquainted with such matters predict that in a few years the Waconda, or Great Spirit Spring, will be one of the most widely-famous and popular in this or any country.

Waukesha, Wisconsin.

THE village of Waukesha is prettily situated in one of the most beautiful and picturesque portions of the attractive State of Wisconsin. Its name has become almost a household word throughout the country, because of the Mineral Springs located there. These springs are among the most celebrated in the land, and their waters are shipped all over the world. But had they never existed Waukesha might have become, naturally, a great summer resort. Its location and beauty, the delightful climate, especially in the later months of summer and fall, its proximity to the lovely lakes, Peewaukee and Oconomonoc, would in any event have made the place one of unusual attractiveness. Its situation upon the Fox River, the principal feeder of the Illinois, adds to its attractions, and so long ago as 1835 it was settled by immigrant pioneers, principally from Indiana and the Eastern States, who were not slow to avail themselves of its admirable facilities for water-power, and those charming beauties of nature which pointed it out as one of the fittest of all possible places in the Northwest for the site of pretty and pleasant homesteads. Since that period, and particularly within the present decade, the village has grown rapidly, until, while retaining all the delightful characteristics of a country life, it presents many of the advantages and conveniences of the city. These have been gained partly from its position, only twenty miles from Milwaukee. About the large springs' hotels parks have been laid out, while the greater number of the private residences and many of the streets are beautifully adorned with shade trees, often meeting and arching overhead, and forming by their branches a protection from sun or shower. The trim lawns and the many-hued and fragrant flowers in the gardens add to the charm of the scene, and altogether in many respects Waukesha presents the appearance of a clean, bright, tidy and flourishing New England village. It has a population of about 5000. Since the discovery of the medicinal properties of the Mineral Springs the place has rapidly grown in favor as a watering-place, and it is now the most popular resort west of the Alleghanies. There are ten springs whose waters are used, of which the Bethesda, Silurian, and Fountain are the best known. In cases of Bright's disease, diabetes, dyspepsia, and all liver and kidney affections, dropsy, gravel, etc., the waters are highly recommended. The hotel accommodations are ample, and there are numerous boarding-houses of various grades and capacities.

Cullisaja Falls, Western North Carolina.

The Health Resorts of the South.

" Knowst thou the land where the lemon trees bloom,
Where the gold orange grows in the green thicket's gloom,
Where the wind ever soft from the blue heaven blows,
And groves are of myrtle and orange and rose?"

URING the last few years all that region of country comprising Northern Georgia, Western North Carolina, portions of South Carolina, and Florida, has been steadily growing in favor among that large class of people who from choice or necessity wish to escape the rigor of a northern climate. The number of those to seek the balmy air of Florida from February to April was greater this year than ever before, and it will be still greater in 1884. To leave snow and frost in the North, with the thermometer near about zero,

Mouth of the Oklawaha.

and be sitting fifty hours later in a shady nook of some Southern piazza with the same thermometer all alive at 86°, a tender, wanton breeze ruffling the bosom of a lake, so far away that no man would ever walk to it—that is if he were as lazy as he ought to be in such an atmosphere—and half-open eyes, catching a glimpse of sun-filled ways, brooded palms, and swaying, blossom-dotted vines, how easy and natural to not only picture the distant spiræa arrayed in white and green, but also to place it in the midst of a velvety lawn, shaded by full-foliaged trees and broken by circles of flowers. It is a translation that makes one wealthy, affable and sociable, and leads him to forget yesterday.

Oddly-shaped Florida is as a rule the first objective point of the traveller south, a country on the edge of the tropics, whose everglades, swamps, and strange rivers bordered by luxuriant vegetation give one an impression of the freaks of nature run wild. Jacksonville, the largest city and capital of Florida, on the St. John's River, about twenty-five miles from the mouth, is modelled on Northern plans, with shady streets crossing at right angles. It is a popular stopping-place and enjoys a busy winter season. The equable temperature is a charm in itself, while there are many pleasant excursions on the river and good views on the fine shell roadways. Those who must have the city as a resort may linger, but to get an idea of tropical scenery one must go by steamboat down the St. John's. Beyond Jacksonville, three hundred miles from its source, the river rolls along, now a stream a half mile wide, and now a lake two or perhaps six miles wide—the low banks netted over with a growth all its own,— a jungle of vines clambering over thickets, and on the hammocks rows of the cotton-wood, the juniper, which sweetens and preserves the waters that glimmer black and deep in the half-

The Suwanee River.

hidden recesses, the red cedar, the sweet gum, the white and black ash, the redo-lent magnolia, the water-oak, and the glistening, richly-dressed palmettoes— and at their feet a maze of shrubbery, amongst which the azalea, the sensitive plant, the sumach, the agave, the nettle, and the poppy are prominent. All these laden thickets are bound together by running arms and tendrils of the fox-grape, while the woodbine and bigno-nia clamber up the great trees and nod in the breeze above. On a jutting cape the heron and the crane pensively yet know-ingly eye the steamer at a safe distance; a splash in the depths beyond marks the spot where a turtle has dropped from a log, or tells that the grinning alligator has taken the hint sent from the chambers of a half dozen re-volvers and "will see you later." Along the river side at intervals are homes lying amongst handsome shade-trees, inviting landing-points, and villages and towns. Some eleven miles above Jacksonville Mulberry Grove is passed, a charming spot for a picnic. A few miles farther on is Mandarin, the winter home of Mrs.

Harriet Beecher-Stowe; then past the high Magnolia Point, and into Green Cove Springs, whose clear green and limpid waters rush out at the rate of three thousand gallons a minute, and are overhung by streamers of gray moss and mistletoe depending from the branches of the encircling oaks; by Picolata, with its old Spanish memories, and on the other side an ancient fort; by Tocoi and several little landings, road stations and orange groves, to Pilatka, the largest town on the way, with a climate made to order and comforts for the invalid. Above Pilatka nature runs wild and frolics everywhere, while the river rolls along in moderation till, just above Welaka and twenty-five miles from Pilatka, it widens into Little Lake George, four miles wide and seven miles long, and then into Lake George, twelve miles wide and eighteen miles long. There is scarcely a lovelier sheet of water in the world than this lake, while the entrance and the exit are all its own. The surface is dotted with islets that

are bowers of vines and flowers, where the creepers run down to the water's edge and repeat their grace and exquisite colors in the mirror below; and here and there are islands under cultivation, with the golden spheres hanging from the midst of the rich, dark green foliage of the orange groves. Around the curves and on the shores the scene is filled by the pelican, the heron, the curlew or the loon, and the flight of brightly plumaged birds, while throughout the clumps of trees the gentle breeze wafts the sweet notes of Southern songsters. Leaving this beautiful spot and passing a succession of forts and landings, Blue Spring, forty miles above, is reached, where mineral waters gush

Wakulla Spring.

out in a strong stream, so clear that the fish can be seen darting about below the shadow of the boat. Further south is Lake Monroe, twelve miles long and five wide, with Mellonville on one side and Enterprise, the head of regular steamboat navigation and a popular resort, on the other.

Twenty-five miles south of Pilatka, opposite Welaka, the Ocklawaha, after flowing three hundred miles, empties into the St. John's. There are no banks to this peculiar stream, which is but a channel through a long series of lakes and cypress swamps. The funny cranky little steamboat puffs into the cypress-shaded opening and winds its way along a river whose only boundaries are blazes on the trunks of the towering trees. The hull bumps against butts of the cypresses and the hidden stumps, and the experience in this line of navigation is odd and original in itself. Curving around the densely wooded turns one may see ahead a mound

8

covered with the tall, slender palmettoes, from whose branches sway gray, fine mosses, and some. times rods and rods of figured patches of swaying, beautifully flowered convolvuluses. Another turn and the boat swings into a green-canopied retreat, where the interlaced and tangled veg. etation overhead shuts out the sunlight and makes a dark cavern below. At the other side on a dead cypress in a solemn row sit a number of buzzards, waiting for nature to add the final touches to a dead alligator before the feast. The swamps on each side abound in birds of many varieties; the water-turkey or snake-bird, hiding his body amongst the foliage, with his long neck and head protruding, or eluding the hunter by dropping into the water and diving to safety; and the white crane, conspicuous and effective in the back-ground. This latter bird has a penchant for juvenile reptiles, and to most it is an event for congratulation when from his sunny bed on a dried palmetto leaf a slimy little imp is picked, to be gently started on the downward slide to the crane's interior. Here too is the paradise of the alligator, which from the wayside winks his piggish eyes as the rifle-ball rattles along his mailed side, or bids farewell to the cypress trees as some experienced sportsman sends a leaden messenger into the vulnerable point in the armor. Sometimes, as the weird little craft bumps along around the "cypress knees," the tangled moss above opens and from the blue sky without the light pours in and flecks the boat in a thousand sunbeams, while the cranes rise up and trail away with flapping wings, the snakes and turtles whisk down to homes beneath the surface, and the brilliant-plumaged paroquets scream as they dart off into the depths. When night comes on blazing pine-knots in the swing-ing cranes on each side of the boat light up the dark channel, and as the rays par-tially illuminate the tall, moss-decked trees around and beyond, strange spectres and grotesque ghosts arise and with supernatural air wave their gaunt arms in beckoning or in despair. One hundred and forty miles from the mouth of the Ocklawaha is the

Grand Chasm, Tugaloo River.

marvellous Silver Spring. Its bosom is a splendid polished mirror, a quarter of a mile wide, its depths as clear as finest crystal for sixty feet down. The steamboat on the surface rests on an inverted fac simile, and every tree, twig, vine and rock is reproduced in the beautiful pool. The floor of this basin is silver sand, studded with curious figures in pale green-tinted lime crystals. A row across the pool is ever to be remembered. Every object that has been dropped into the water by preceding visitors lies in the silver setting, a rich emerald gem. At one place a barely discernible bubbling points out the spot from which the water gushes out, thousands of gallons, every moment. A stone dropped toward the slight ledge of limestone rock twenty-five feet below, is suddenly thrown in a curved line nearly to the surface by the rush of the spring from under the rock. A turn of the boat around the corner into the sunlight and one can scarcely believe that there is anything between his craft and the sharp silhouette on the sands below. The river may be followed for some ninety miles farther, past some picturesque and lovely lakes, into the remote wilderness, where frost rarely penetrates, and sugar-cane tassels.

Toccoa Falls, Northern Georgia.

Extending southward, on the east side of the peninsula, for nearly a hundred and fifty miles from the lower end of Mosquito Inlet, and separated from the ocean by a narrow strip of land, runs Indian River, a long lagoon or arm of the sea. For thirty miles the St. John's and the Indian River run in parallel lines, ten miles apart. The water of the lagoon is salt, and is rich in fish of every kind, including the delicious pompano, and abounds with rare oysters and turtles. A belt of evergreen woods marks the eastern side, and tempers the winds to the rheumatic and consumptive, who find in this country the needed quiet and tonic. In parts of the Indian country bears and deer may be hunted, while an abundance of smaller game give ample employment to the sportsman. The region on the western side is fertile and but awaits enterprise and capital to make it bloom.

One of the most developed parts of the State is " Middle Florida," the section surrounding the capital, Tallahassee, one of the most pleasant cities of the South, resting on an elevation and fanned by the breezes from the Gulf. About fifteen miles from the city is one of the chief wonders of the State, Wakulla Spring, which sends off a river from its single outburst.   The experience of Silver Spring may be here renewed,—the same lime-impregnated, thrillingly transparent water, and the same mosaics of graduated green hues.   The basin is narrower than that of Silver Spring, but in one particular more impressive, being one hundred and six feet deep.   Fifty feet below the surface one may see a great ledge of white rock, from beneath

which the fish swim out.   You look down past the upper part of this ledge, down, down through the miraculous lymph, which impresses you at once as an abstraction and as a concrete sub-stance, to the white concave bottom, where you can plainly see a sort of trouble in the ground. As the water bursts from its mysterious channel one feels more than ever that sensation of depth itself wrought into a substantial embodi-ment.   Proceeding from this spring, the Wakulla pours into St. Mark's River, a mysterious and picturesque stream, which at the Natural Bridge disappears into the earth for the space of fifty feet.   In "West Florida" there are many points of interest and spots where the invalid may re-cuperate.   The principal cities are Pensacola and Appalachicola, the former situated on Pensacola Bay, a body of water of some two hundred square miles area, and the latter at the point where the river of the same name empties into Appalachicola Bay.   The climate and position of both places are all that could be desired, while there are picturesque ruins and forts about the former city.   At the southern extremity of Florida, on an island of the same name, is

Key West, next to Jacksonville the largest city in Florida.   The island is seven miles long, from one to two miles wide, and eleven feet above the sea.   It is interest-ing as being of coral formation, a fact that modifies the mode of living in many ways.   •

In various parts of the South there are cities which have a national reputation as winter health resorts.   Foremost amongst these is St. Augustine, the oldest European settlement in

Cullisaja Falls, from the Chasm.

the United States.   Its history is interesting and romantic, carrying one back to the Middle Ages and the times when Spanish cavaliers ventured across the great deep in search for Eldorado and the fountain of eternal youth.   "The aspect of St. Augustine," says Mrs. Harriet Beecher Stowe, " is quaint and strange, in harmony with its romantic history.   It has no pretensions to architectural beauty, and yet it is impressive from its unlikeness to anything else in America. It is as if some little, old, dead-alive Spanish town, with its fort and gateway and Moorish bell-towers had broken loose, floated over here and got stranded on a sandbank. Here you see the shovel-hats and black gowns of priests; the convent with gliding figures of nuns; and in the narrow, crooked streets meet dark-browed people, with great Spanish eyes and coal-black hair. The current of life here has the indolent, dreamy stillness that characterizes life in old Spain. In Spain, when you ask a man to do anything, instead of answering as we do, 'In a minute,' the invariable reply is, 'In an hour;' and the growth and progress of St. Augustine have been according. There it stands alone, isolated, connected by no good roads or navigation with the busy living world." The streets are narrow, and consequently in that warm climate shaded and draughty. A vehicle is rarely seen on the streets, and the shifting sand lies over the broken shell-concrete that formerly paved the way. On each side are old Spanish houses, built of coquina stone, a peculiar conglomeration of fine shells and sand, which are first stuccoed and then whitewashed, while the quaint hanging balconies of the second stories almost touch from side to side. In the newer parts of the city are modern dwellings and hotels, and many elegant winter villas. At the northeast end of the town and fronting to the sea is the old fort of San Marco, built of coquina. It was begun in 1656, and according to the inscription, handsomely cut in the stone under

Canyon of the Cataleuche.

the arms of Spain, was finished one hundred years thereafter. It is a royal old pile. Its castellated battlements, its heavy bastions guarded by frowning guns, its lofty and imposing sally-ports encircled by the royal Spanish arms, its moat, drawbridge, and portcullis, its round and carved sentry-boxes at each prominent parapet-angle, its high lookout tower, and its time-marked and moss-grown, massive walls, make a fit exterior for the heavy casemates within, the dark passages, gloomy vaults, and hidden dungeons, the ruined Romish chapel, with its ornate portico, and inner altar, and holy-water niches; and one as he rambles through this relic of departed ages dreams of knights in armor charging up to the walls, hears the solemn chanting of friars, and the rude laughter of the rough soldiers, and in the depths of the dungeons looks for a decaying skeleton, and listens for the clanking of rusty chains. The entire ocean-front of the city is protected by the Sea Wall, running from the water battery of the fort southward a mile. A delightful moonlight promenade is along the four feet wide granite coping of this wall. The Plaza de la Constitucion, a fine public square, with seats about it, lies in the centre of the town. Fronting on the square among other buildings is the striking old Catholic Cathedral. It has a quaint Moorish belfry, with four bells dating back

Sugar Fork Falls.

to 1682, and a clock so placed as to form a perfect cross. There are a number of other interesting and imposing buildings in the old city, and the lover of antiquities may find ample opportunity to gratify his taste. All about the city are pleasant points to which excursions are made, and a charming drive leads out St. George Street through the city gate, a relic of the old Spanish wall, with its carved towers, loop-holes and sentry boxes, forming a picturesque structure. The climate is that which prevails throughout the favored State, which, though ten degrees lower than Southern Italy, is so influenced by the counter-currents in the ocean as to maintain an equable temperature no higher than that of the country across the sea.

A city with a history is Charleston, South Carolina. Its name has figured in the annals of every war, from the proud day that saw the British balls sink in the palmetto logs of Fort Moultrie and the hostile ships sail away defeated, to the sad hour when Fort Sumter was fired upon. The bright sunny winters, and a yearly mean temperature of sixty-six degrees draw many visitors,—the sick, who find here a delightful climate and the needed comforts of a city, and the gay, who make the year round a perpetual spring. There are many interesting drives along the Ashley and the Cooper rivers, and around Sullivan's Island. In the suburbs of the city are a number of old planters' houses, Drayton Hall, Middleton Place, and The Oaks being especially notable, with their elegant lawns and the evidences of former splendor, on which War laid a rude and unsparing hand.   Magnolia Cemetery is lovely in shrubbery and flowers, and holds the remains of several distinguished men.   The town itself has imposing public buildings, and some old and attractive churches,—amongst others the venerable St. Michael's, built in 1752, with tall belfry, holding sweet chimes; and St. Philip's, by whose walls John C. Calhoun is buried.   Interesting trips may be made to the rich "Phosphate Mines," along the Ashley and Bull rivers, and to the forts and islands in the harbor.

The climate and conditions of Savannah and Augusta are very similar to those of Charles-ton, and the three cities are much resorted to by consumptives and other invalids who desire to remain within the region of postal delivery.   Savannah occupies a bluff on the river some forty feet high, running back to and including a portion of a plateau in the rear of the city.

McDowell's Hill, French Broad River.

There are twenty-four parks within the limits, and the whole city abounds in trees, shrubbery, and flower gardens, that bloom throughout the year.   In the southern section is Forsyth Park, having an area of about forty acres.   A notable object is the Pulaski Monument.   There are lovely drives leading to White Bluff, Montgomery, Beaulieu, Isle of Hope, and Thunderbolt out on "The Salts." The finest drive, and one of the most picturesque in the country, is that to Bonaventure Cemetery, on Warsaw River, and about four miles from the city.   It was once the residence of the Tatnalls, an old English family.   The beautiful city of Augusta, the third in population in Georgia, lies at the head of navigation on the Savannah River, on a broad and picturesque sweep.   The city is handsomely laid out, and is famous for its fine avenues, the principal one, Green Street, being one hundred and sixty-eight feet wide, having a grass space in the centre lined on each side by a row of shade trees. The Fair Grounds, occupying forty-seven acres, are just outside the city. On the high hills some three miles from Augusta is Summerville, a handsome suburb, where many Northerners own villas. Horse cars connect with the city, and many find this quiet and attractive little place preferable to the more populous points. There is a town of the same name some twenty-two miles from Charleston, on the South Carolina Railroad. On the line of the same road, sixteen miles from Augusta, and one hundred and twenty from Charleston, on a sandy plateau six or seven hundred feet above the sea, lies Aiken, the most frequented winter resort in the United States. There is scarcely any soil, and everywhere in the town is clean white sand. The air of Aiken is drier than that of any other prominent Southern resort, and in the matter of equability of temperature is surpassed only by San Diego. The mean temperature is as follows: Spring, 63.4°; summer, 79.1°; autumn, 63.70°; winter, 46.4°; for the year, 63.1½°. The average rainfall, as follows: Spring, 11.97 inches; summer, 13.89; autumn, 7.34; winter, 7.16; for the year, 40.36 inches. These data are from the record of 1870. The surroundings of Aiken are as tranquil as the temperature is equable. Straight vistas run out over the sands through the sombre pine woods that encircle the town. There are no hills to climb, no falls to visit, no commanding views in the neighborhood one *must* see. The houses are the wide-porticoed typical Southern houses, with a chimney sustaining each side, a sunny, open yard, flowering vines, piles of roses, and the general hospitable, welcoming air. The little negroes drive a thriving trade in the sale of the fifty varieties of sand that are found here, ranging in color from green to brownish red, with now and then traces of blue. There is no business but that of

Cascades near Warm Springs.

entertaining the guests, and no noise of traffic. Everything is peaceful and quiet, and the chief charm of life comes from the beauty of the clear winter days.

The patient having in his sojourn in sunny climes become convalescent, will enjoy the exercise and tonic of a ramble in a mountainous country, and nowhere in the Appalachian system could he find a better field than in the northern districts of South Carolina, North Georgia, and Western North Carolina. One of the remarkable wonders of South Carolina is Table Mountain, (4300 feet high,) with a barricade of perpendicular cliffs, one thousand feet high on one side, which present a grand appearance from the wooded glens below. From the summit of the mountain may be had a fine view of the conspicuous Cæsar's Head. The Falls of Slicking, a marvellous series of cascades and rapids, lie at the base of Table Mountain. Down the declivity two streams rush, joining at a point called the "Trunk," from which a most charming view of Cæsar's Head, Bald Mountain, Pinnacle Rock, and adjacent peaks may be obtained. The two streams fall over seventy feet at this point into a glen, wild and picturesque as any on the continent. The pretty mountain stream, the Keowee, runs through the rare little Jocasse Valley, and varies its course in a leap at the White Water Cataracts.

From Clarksville, Habersham County, Georgia, there are several roads to the mountain country. A few miles from this town are the celebrated Toccoa Falls, where a stream comes through a chasm in the hills to tumble perpendicularly over a great rock from a height of one hundred and eighty-five feet, and upon reaching the bottom is dispersed in mist, which, visible to the eye against the dark background, waves to and fro in a weird manner. Tallulah Falls are distant twelve miles from Clarksville. Tallulah "the Terrible," a large stream, here breaks through the last obstacle in its eastward course, and for two miles, through a gorge of twelve hundred feet in depth and of unsurpassed grandeur, is dashed over deep falls, over great rocks, and broken into cascades in the wildest manner. It requires steady nerves and strong muscles to visit the different points of interest along the edge of the chasm, or to scramble down its deep and

Watauga Falls.

rugged face to behold the mad struggle of the troubled river. The Falls are made up of numerous cataracts,—the Lodore, the Tempesta, the Oceana, and the Serpentine among others. Two notable points are the Pulpit and the Trysting Rock; and the wild chasm is filled with imposing granite walls and boulders, leaping waters and deep gorges, which give it high rank in the scenery of our country. The Valley of Naobochee, the Falls of Eastatotia and Amicalolah, and Nickojack Cave are other points to be visited. A ride along the Richmond and Danville Railroad gives one an excellent idea of the scenery of the country.

The mountain country just gone over, however, pales before the grand and impressive "Land of the Sky," in Western North Carolina. The Blue Ridge and the Great Smoky, two great mountain chains, encircle a plateau two hundred and fifty miles long and fifty broad, which is crossed by four transverse ranges, the Black, the Balsam, the Cullowhee, and the Nantsahala. The Black Mountains are the most famous, and include Mount Mitchell, the loftiest summit east of the Mississippi. The centre from which the route into the mountains diverge is Asheville, 2250 feet above the sea, and from this pretty town, with a charming climate, one may visit the Linville Gorge, with masses of broken, tumbled granite rocks

A Glimpse of French Broad River.

and beetling cliffs, 2000 feet high, with a river churning and dashing the ragged way, and traverse the picturesque Swannanoa Gap, around Cassair's Head into Cashier Valley, to climb the noble Whiteside Mountain, probably the most striking peak in the State. Five thousand feet high, its face is a tremendous curve of white rock, eighteen hundred feet high and two miles long. The face, at a distance apparently smooth, is in reality worn and eroded, having many peculiar recesses, amongst others the Devil's Supreme Court-house. One of the loveliest glens is the famous Hickory Nut Gap, through which the French Broad River runs past many a curiously carved pillar. The climbing of Mounts Pisgah and Mitchell is a matter of course, and the scenes from the summits repay the toil of the ascent. At the top of the latter lies buried in his monument Professor Mitchell, in honor of whom the mountain is named. A cairn of stones, to which each visitor adds his mite, is slowly building over the last resting-place of him who, while exploring the great mountain, was dashed to death in one of its many chasms. The Cullisaja, the Sugar Fork, and the Wautauga Falls are all charming cascades, and some of the most beautiful scenery is found along the course of the Richmond and Danville Road, by the Cañon of the Catalouche, and on the shores of the French Broad River.

No one intending to travel South should expect to find a Paradise, and be enraptured everywhere. There are times and conditions that make many a trip sadly disappointing; but "to him who in the love of Nature" goes abroad, this section of our great land will afford the widest opportunities to view her in some of her sweetest and most charming moods. The greatest mistake may be made by invalids who seek resorts indiscriminately, without regard to constitution or circumstances; and many poor creatures in the last stages of fatal disease are torn from the comforts of a home to die amidst strangers. In all cases a competent physician should be consulted. The cities enumerated in the body of this article are, however, helpful in almost all cases requiring pure air, a steady temperature, and peaceful, quieting influences.

Lakes and Rivers.

" How beautiful the water is !
To me 'tis wondrous fair—
No spot can ever lonely be
If water sparkle there :
It hath a thousand tongues of mirth,
Of grandeur or delight,
And every heart is gladder made
When water greets the sight."

To nearly all mankind water hath its charms. The very mention of a beautiful lake, with its settings of mountain steeps, woods and rocks, or a deep-running, winding river, with its banks of verdure and flowers and shady nooks, is a suggestion of beautiful thoughts and pleasurable emotions. Moonlight on the water describes the very essence of romance. To the heat-oppressed inhabitants of the parched and dusty city, in July and August, the thought of being embowered in some cool retreat by the side of a lovely and picturesque lake or river is a picture of perfect comfort and earthly bliss. Of these retreats and the beauteous waters which make them attractive our own land has a bountiful supply. First of all is the great chain of lake son our northern boundary, which clasp hands and extend from Minnesota to the shores of the Atlantic. These five sister lakes—Superior, Michigan, Huron, Erie, and Ontario—which pour their waters through the St. Lawrence to the ocean, are the most extensive inland seas in the world, and each has its distinguishing characteristics of scenery and suggestion. They all abound in features of interest to the tourist, and many delightful summer resorts are located on their borders. Lake Superior, the largest and most mysterious of the chain, whose waters are daily churned into a foam by the paddle-wheels of steamboats, is only half explored in its northern shores, and strange and fairy-like tales are daily told by fur-traders and hunters of gold and silver, rubies and amethysts, copper and tin, to be found in the trackless regions washed by its waters. The celebrated Pictured Rocks, stretching from Munesing Harbor eastward along the southern coast, are among the wonders of the New World. Lake Michigan is perhaps the most beautiful of the series. Nothing is more soothing than the soft air wafted over its cool, sea-green waters ; nothing more delightful than a sight of its beautiful islands, shifting fogs, and unsurpassed Straits of Mackinaw. The Island of Mackinaw, a spot sacred to the Indians of the lakes, is scarcely lacking in any of the beauty or interest to be found in the Yosemite or Yellowstone national pleasure-grounds. Perhaps the most romantic of the chain is the deep blue Huron, with its wild shores and far-stretching woodland solitudes. Sault Ste. Marie, connecting it with Superior, is but little inferior in beauty to Mackinaw. No place in our country is so fraught with incidents relating to our national colonial life as Lake Erie. The spirit of Pontiac haunts the mouth of the Detroit River. On the shores of the lake every tree in the woods, as the winds sigh through its branches, whispers the name of Tecumseh, and his farewell to his British allies, with his declaration to lay his bones on the battle-field without retreating. The renowned resort of Put-in-Bay reminds the world of the immortal Perry and his famous dispatch, " We have met the enemy and they are ours ;'' and call to mind the dying words of Captain Lawrence, " Don't give up the ship,'' which Perry inscribed upon a flag, flung to the breeze from the mast-head of his vessel. Anthony Wayne's laconic field-order, " Charge the d——d rascals,'' is remembered at the pronunciation of the name of Presque Isle. Charming and sublime Ontario, though in a degree dulled by the sublimity of Niagara Falls and the picturesque loveliness

(116)

of the Thousand Islands, is surrounded by natural scenery of surpassing beauty, and forms a fitting climax to this sublime and beautiful series of great inland seas.

Lake Geneva, Wisconsin.

The Northwest has, within the past ten years, developed many beautiful spots wherein the warm months of summer can be pleasantly passed, and where health and strength may be restored to the invalid. The States of Minnesota and Wisconsin have led the van in the number of these places offering attractions of scenery, climate, recreation and amusement. The resorts of Wisconsin, because they are so easily reached from the great centre—Chicago—and because they furnish all the attractions to be desired by the most fastidious, have become more noted than those in any other section, and many of them are rapidly acquiring a reputation that may well be envied by the older places of resort in the East. For all the many reasons that have made Wisconsin popular to the summer saunterer, has Lake Geneva taken the lead of Western resorts. It presents all the advantages that could be asked for as regards climate, scenery, good society, and means of recreation and amusement. Lake Geneva lies forty-four miles southwest of Milwaukee, and sixty-two miles northwest of Chicago, as a bird flies

The lake is nine miles long by about two wide. Its depth is very great, and in places no bottom has been found. It has no inlet, but is supplied entirely by pure spring water gushing from the hillsides along its picturesque shores. No slough or malarial pools are found about the lake, and a weed has never been seen in the lake, and no insects and flies, so common in weedy and marshy lakes, are here found. Its waters are so clear and transparent that the bottom, as well as fish and other objects, can easily be seen at a depth

Summer Residence of N. K. Fairbank, of Chicago.

of thirty-five feet. Nothing but charming pebble and boulder shore-line is to be seen, and in places these boulders line the gracefully curving shore for miles in length, lying as neatly as if a master mason had fitted them in the line of beauty. The scenery is nowhere wild; it is such as painters love to delineate and lovers of art delight to view. The ever-changing hue of the waters from deep blue to ocean green, is, in itself, an enjoyable study, even to old acquaintances. Over the lake itself, in the last fallen hours of the day, hangs a curious purple-gray, making the freshly-painted boats and wooded banks seem like the pictures in a dream. In the shadows of the grove springs

> " That delicate forest flower,
> With scented breath and look so like a smile,
> The moss-clad violet, fragrant and concealed,
> Like hidden charity."

Lake Geneva is different from other Western resorts in that it is distinctly a family.watering-place. Its visitors come in June and stay until October. The entire twenty-five miles of its beautiful shore is occupied by the residences of wealthy citizens of Chicago and St. Louis, for the most part. The majority of these houses are expensive and elegant, and have been built at an expenditure of from twenty-five thousand to a quarter of a million dollars. The amount of money invested in these summer homes will foot up among the millions, and they give a character and prestige to Lake Geneva possessed by no other western resort.

The lake has been artificially stocked with all kinds of game fish, and the fishing in the proper season is excellent and free to all. It is particularly. noted as being the home of the "cisco," a species of white-fish found nowhere else in the world. In the full of the moon in June, these fish come to the surface, and for a few days thousands of them are caught with a hook. They then disappear and none have ever been seen during the balance of the year. During the "run" of the fish, the air is filled with a peculiar fly, which disappears with the ciscoes, not to be again seen till the next year. The fish caught from the lake are very gamy, particularly the black bass. Dozens of small lakes fairly full of fish are located within a few miles from Lake Geneva, and

Summer Residence of Julian S. Rumsey.

are easily reached in an hour's drive. There are five public steam.rs on the lake, which can carry from fifty to six hundred passengers. Besides these, many of the summer residents own steam yachts, which have the reputation of being the finest of their size in the country. The average is 75 feet long and cost from fifteen to thirty thousand dollars. Speed is the chief requisite of the private yachts. Sail yachts are numerous and sailing is very much enjoyed, many regattas taking place during the season. The hotel and boarding-house accommodations are ample and reasonable. The place has always been free from the charges of extortion often heard in connection with pleasure resorts. It has been called "the New-port of the West," and it is to Chicago what Newport is to New York.

Lake George, New York.

THIS unrivalled gem of American lakes is found at the southwestern margin of the great Adirondack Wilderness, thirty-one miles north of Saratoga, and two hundred and eleven miles from New York. It is thirty-four miles long, running north and south, and varies from two to four miles in width. The lake is literally embowered in beautifully-wooded hills, which in many instances rise abruptly from its margin and attain an altitude of more than two thousand feet. Its pellucid waters come entirely from the mountain brooks, and springs coming up from the bottom of the lake. Lake George is studded with many small islands— one for each day in the year, with one accommodating little fellow, which is understood to be held in reserve for the 29th of February. Lake George is made interesting by history and legend as well as by the great beauty of its scenery, for which it is renowned throughout the the world. The battle of Lake George is a prominent event in our colonial history ; and the inhabitants, Hawkeye, Chin-gach-cook, Uncas, Alice, and Cora Munro, the creations of the

Lake George—The Narrows.
FROM STODDARD'S GUIDE TO LAKE GEORGE.

genius of our great novelist, Cooper, will never be dispossessed of it, but will ever remain associated with it in the minds of all lovers of American literature. There can be no more charming excursion than a passage up and down this American Como affords. The wild, picturesque shores, the pretty little bays, the fascinating islands, the soft glamour of the water, and the towering hills, make an enchanting panorama. Caldwell, the principal resort, is situated at the head of the lake, and the village of Baldwin at the foot, where it empties into Lake Champlain. Across the point of the Lake from Caldwell is Crosbyside, quite a popular resort. Just east of Caldwell, and commanding the most beautiful view of the lake and its surroundings, is the far-famed Fort William Henry Hotel. Numerous other resorts are located along the shores of the lake, on the waters of which a regular line of steamboats is run, making three trips daily between Caldwell and Baldwin, touching at all of the intermediate landings. Caldwell may be reached *via* Delaware and Hudson Canal Railroad and its connections.

Otsego·Lake, New York.

This beautiful lake, situated in Otsego County, New York, is about nine miles long and one to one and a half mile wide. J. Fenimore Cooper, the novelist, in his " Deer-slayer," thus describes the lake and surrounding hills: "On a level with the point lay a broad sheet of water, so placid and limpid that it resembled a bed of the pure mountain atmosphere compressed into a setting of hills and woods. At its northern end it was bounded by an isolated mountain; lower land falling off east and west, gracefully relieving the sweep of the outline; still the character of the country was mountainous: high hills or low mountains rising abruptly from the water on quite nine-tenths of its circuit. But the most striking peculiarities of the scene were its solemn solitude and sweet repose. On all sides, wherever the eye turned, nothing met it but the mirror-like surface of the lake, and the dense setting

Otsego Lake, New York.

of woods. So rich and fleecy were the outlines of the forest that the whole visible earth, from the rounded mountain-top to the water's edge, presented one unvaried hue of unbroken verdure." A recent writer says: "The same points still exist which Leather Stocking saw. There is the same beauty of verdure along the hills, and the sun still glints as brightly as then the ripples of the clear water." The scenery along the shores is extremely picturesque, and in the transparent waters there is found an abundance of fish. The whole region is full of interest because of the creations of Cooper's genius, and his romances have a new zest and beauty when read amid the scenes which inspired them. Cooperstown, situated at the south end of the lake, is the principal resort in this section of the State. It is beautifully situated high up in the mountains in the midst of delightful scenery, and has a clear, bracing atmosphere. The old Cooper mansion where J. Fenimore Cooper lived was burned in 1854. The site is always visited by tourists, however, as is also the Tomb of Cooper, Cooper's Monument, Leather-Stocking Cave, Leather-Stocking Falls, and a hundred other points of interest in the vicinity. Two small steamers ply on the lake, touching at all points of interest along the shores, affording opportunities for delightful excursions. It is claimed that hay-fever is unknown here, and that victims of the disease always find relief. Cooperstown is reached *via* the Albany and Susquehanna and the Cooperstown and Susquehanna Valley Railroad. Distance from Albany, ninety-one miles.

Lake Winnepesaukee, New Hampshire.

THIS beautiful sheet of water, lying in the central-eastern part of New Hampshire, has located on its shores some of the most attractive summer resorts in New England. The waters of the lake cover an area of over 70 square miles. It is quite irregular in outline, and very shallow, at no point attaining a depth of over two hundred feet. There are as many islands in the lake as there are days in the year. It is supposed that the bottom of the lake contains many large springs, as the streams which flow into it are altogether incompetent to create the great mass of water which it contains. "There may be," says Bartol, "lakes in Tyrol and Switzerland which, in particular effects, exceed the charms of any in the western world; but in that wedding of the land with the water, in which one is perpetually approaching and retreating from the other, nothing can be held to surpass, if to match, our Winnepesaukee." From the shore the range of vision is soon stopped by the islands, which can hardly be separated from each other in the dim distance, but from the summit of any one of the numerous mountains which surround the lake the whole extent of its surface is spread out like a map and glitters in the sunlight like a sheet of crystal sprinkled with emeralds. Centre Harbor, at the head of the long North Bay of the lake is one of the chief summer resorts of this region. It is a small hamlet occupying an excellent position for studying and appreciating the beauties of the lake. The steamers "Lady of the Lake" and "Mt. Washington" touch at this point three or four times daily, and stages leave every afternoon for Moultonborough and West Ossipee. Wolfborough, the terminus of a branch of the Eastern Railroad, is the largest village on Lake Winnepesaukee. It is prettily situated at the foot of Wolfborough Bay, the most easterly projection of the lake, and commands a view of the entire bay and part of the open lake. It is a popular and greatly-frequented resort. The lake steamers touch at Wolfborough several times daily. Alton Bay, at the southern extremity of the lake, and the terminus of the Dover and Winnepesaukee branch of the Boston and Maine Railroad, has good fishing and offers good views of the White Mountains, but is less popular as a resort than most of the other villages on the shores of the lake. This place is also one of the landings of the "Lady of the Lake" and "Mt. Washington." Other pretty but less frequented resorts, as Weir's, Meredith, Moultonborough and others, afford fine views of lake and mountain scenery, and are splendid starting-points for numerous excursions by water, stage, carriage, or for foot tours. They are all reached by stage and the lake steamers, and some of them by rail. Edward Everett, in speaking of a trip by steamer from Weir's Landing to Centre Harbor said that he had been something of a traveller in our own country, and in Europe had seen all that was most attractive, but his eye had yet to rest upon a lovelier scene than that which smiled around him as he sailed from Weir's to Centre Harbor.

Lake Memphremagog, Vt.

AWAY in Northern New England, nestling among the mountains, partly in Vermont and partly in Canada, is lovely Lake Memphremagog, declared by many enthusiastic tourists to be equal in beauty to Lake George. It is thirty miles long and about two miles wide, and extends in a curve, following the mountain range, from Coventry, Vt., to Magog, Canada. Its clear, pure waters are the home of many large speckled trout, which invite the disciple of Isaac Walton to sojourn here and try his skill. Its shores are of varying character and outline —now high and rugged cliffs wall in the waters, again thickly wooded hills guard the shores, and anon the sweet green meadows stretch out and touch the margin of the quiet lake with their pebbly, sandy margins. Numerous gems of islands bedeck the lake, many of which are cultivated, and some (chief among which is Tea-Table Island,) are devoted entirely to pleasure.

A trip up and down the lake affords the tourist a continual succession of beautiful scenes. On Lake Memphremagog, as at most lake resorts, the mountains only furnish a background for the charming lake scenery itself. Newport, Vt., at the head of the lake, is, perhaps, the principal

Lake Memphremagog, Vermont

resort in this region, though the tourist will see in a sail down the lake many pleasant summer hotels show their low white buildings on the shore, and from time to time pretty villas rising among the embowering trees. Newport, three hundred and sixty-five miles from New York, and two hundred and thirty from Boston, may be reached *via* the Passumsic and connecting railways.

Moosehead Lake, Maine.

AMONG the northern hills, on the verge of the great Maine forest, stretching away into a wild and yet mostly uninhabited region, is Moosehead Lake, the largest sheet of water in the Pine-Tree State. It is ten hundred and twenty-three feet above the level of the sea, into which, by way of the Kennebec River, it pours its waters. Its shores are of irregular outline, and its waters deep, clear, and cold, furnishing ample occupation to the angler in their stores of trout and other fish. Vast numbers of game, including deer and moose, still frequent the densely wooded boundaries of the lake. Owing to these facts it has in recent years possessed a high reputation among tourists and sportsmen. Greenville, a small village on the southern extremity, is the only permanent settlement on the borders of the lake, though several summer hotels are located in the vicinity. A small steamboat plies daily between Greenville and Mount Kineo, a prominent summer resort and favorite stopping-place on the east shore. The steamer also makes pleasure trips to the northern end of the lake, the passage to which affords a panoramic succession of fine scenery. The most striking and imposing scene along the shores is Mount Kineo, which rises precipitously from the water to a height of over six hundred feet. The summit of the mountain, which is easily reached from the hotel located at its base, reveals a magnificent picture of forest, mountain, and water. From no point can so fine a view be obtained of grand old Mount Katahdin as from the top of Mount Kineo. The chief drawback to a visit to this or any portion of the Maine woods is the blackfly, which from the middle of June to the first or middle of August are "masters of the situation," though at other portions of the season they are not troublesome. Greenville may be reached by stage from Skowhegan, Dexter Station, or Guilford, all of which places have railway connections.

Lake Winnepesaukee, New Hampshire.

Lake Champlain, New York.

FEW places in America have so many historical and romantic associations as Lake Champlain. It was known to the Hurons, Algonquins, Iroquois, and other tribes of Indians, as the "Gate of the Country," and was the centre of as many striking events in their rude warfare as it afterward proved when the French, English, and Americans expended life and treasure in struggles for its possession. Crown Point, Fort Ticonderoga, (which still remains, a most picturesque old ruin,) and many other places along its borders are invested with especial interest connected with our colonial history, the revolutionary war, and the war of 1812. Lake Champlain was the arena of one of the most brilliant naval feats in the last-named war—the defeat and capture of nearly the entire British fleet by Commodore McDonough. In this naval battle Commodore McDonough had fourteen vessels, eighty-six

Lake Champlain, New York.

guns, and four hundred and fifty men, while Captain Downie, who commanded the British fleet, had sixteen vessels, ninety-five guns, and one thousand men. The American commander's success was due entirely to the skilful management of his vessels, and the bravery of his men. Since that event the waters of Lake Champlain have been unruffled by strife. Fleets still sail over the lake, but the ships bear charmed and delighted tourists; armies still invade the surrounding territory and scale the mountain heights on the shore, but they are armies of enraptured and gratified summer visitors and health-seekers. Lake Champlain lies between the Green Mountains on the east and the Adirondacks on the west, on the border-line between Vermont and New York. It is one hundred and twenty-six miles long, of varying width, and of very irregular shape, beginning in a series of long, crooked reaches, so narrow that it would be difficult or impossible to turn an ordinary steamboat in them, and widening above Ticonderoga, until at a point near Burlington, Vermont, it attains a width of ten miles. While the lake is surrounded with mountain ranges which stretch far away on either hand, there is an absence of steep cliffs directly on the water, a general characteristic of the shores of our northern lakes. Broad acres of beautiful meadow and farm lands are frequently seen sloping down to the shores, upon which smiling homes are located, and where peace and plenty have their abode. This beautiful little inland sea, with its sister, Lake George, will always remain among the most favored goals of summer pilgrimage. A trip on Lake Champlain by the elegant and commodious passenger-steamers which ply between the different places and points of interest, is indescribably delightful. Burlington, Vermont, situated on the eastern shore of the lake, has in recent years become quite a popular headquarters for tourists whose objective points are the Green Mountains, the Adirondacks, and places of interest along the lake. Tourists having "done" the White and Green Mountains, and proposing a trip through the Adirondacks, will have the pleasure of their "vacation" greatly heightened by tarrying a few days at Burlington and indulging in some of the many delightful excursions that may be taken from here. From Burlington the visitor may take the boat, cross to Port Kent, and go down to Plattsburgh, from whence, by the New York and Canada Railroad, Ausable chasm may be visited. Plattsburgh is one of the favorite entrances to the Adirondack region. Probably one of the finest tours for its length in the world, which includes a sail over Lakes George and Champlain, may be taken from New York up the unrivalled Hudson to Albany, through Saratoga, over the lakes, through the whirling rapids, and past the Thousand Islands of the far-famed St. Lawrence, on the broad bosom of Ontario, to Niagara Falls, and back to the metropolis through the varied beauties of the Empire State.

Chautauqua Lake, New York.

AMONG the many beautiful inland lakes in the Empire State none deserves a wider notice than that of Chautauqua—a body of purest water, seven hundred and eighty feet above Lake Erie and only eleven miles distant—with the thriving village of Jamestown at its southerly end and Maysville at its northerly. Its name, of Indian origin, was early given to the territory now famed as Chautauqua County. Years ago it became known that this lake, then supposed to be the highest navigable water on the continent, was surrounded by a beautiful region of country—with a summer climate pure, and healthful, and invigorating—and year by year many visitors, leaving the crowded cities, had gone into camp on the beautiful banks of this delightful lake, whiling away the restful days in capturing pickerel, a fish that grows to immense proportions in its crystal waters. As time wore on it was discovered that the healthful atmosphere of this region was a panacea and sure cure and safe protection against "hay fever," and multitudes flocked hither to find exemption from this fearful malady. In 1873 Rev. John H. Vincent, D.D., a man of world-wide fame, now known as the "Bishop of Chautauqua," camped on the shores of this beautiful lake—for a respite from labor and weary toil. He came for rest, but while here he carved out the "Chautauqua idea" of elevating humanity by furnishing them with healthful recreation under moral and Christianizing influences. In the summer of 1874 he initiated and called together the first annual "Assembly," which was held for two weeks in a beautiful grove near the foot of the lake at Fair Point, now known all over the world as Chautauqua. This first experiment meeting with such unexpected success, an organization was effected and a charter with ample powers was obtained from the State of New York. Lands were purchased, and the foundations were laid for a growth that has ripened into a summer resort with a thousand elegant cottage homes—with a grand hotel costing one hundred thousand dollars, a city with elegant walks and parks, lit with electric lights, with all the comforts and conveniences of Cape May, Long Branch, and

Saratoga. Year by year these summer gatherings have increased, and many families from the South own cottages to which these families come to spend the entire summer. The organization now owns one hundred and twenty-five acres of land and has a plant of two hundred and fifty thousand dollars. It is known as a " University in the Woods," has a Hall of Philosophy, a Children's Temple, an Alumni Hall, a Park of Palestine, Model of Jerusalem, a Tabernacle in the Wilderness, section of the Pyramid Cheops, an Oriental Museum, an immense auditorium and an amphitheatre seating seventy-five hundred, etc., etc. It has a school of languages, its Chautauqua library and scientific circle, with circles and members all

Chautauqua Lake, New York.

over the world. Its annual Sunday-school assembly collects together thousands of the brightest and foremost thinkers in this and other lands. It is safe to say that one hundred thousand persons visit this charming resort each summer. Fully a dozen steamboats ply upon the lake, with bands playing and banners flying, making summer a grand hey-day of pleasure. Chautauqua Point, Griffith's, Lake View, Mayville, and Jamestown have large hotels, which are filled to overflowing during the gay season. Chautauqua is reached from Washington, Baltimore and Philadelphia via the Pennsylvania Railroad, connecting with the Buffalo and Western at Corry or via same to Buffalo. From New York take New York Central to Buffalo, or the Erie to Mayville.

Greenwood Lake, New York.

In Orange County, New York, near the line of the Erie Railway, nearly hidden by rugged mountains, and surrounded by scenery, which for grandeur and picturesque loveliness is unsurpassed, is little Greenwood Lake. It is ten miles in length and scarcely one mile in width. The forests, with which the surrounding mountains are clothed, reach to the banks of the lake, the waters of which are cold and deep, and so beautifully clear that fish, with which

it is plentifully supplied, may be seen many feet below the surface. The pleasure here afforded tourists for boating is unsurpassed. A small steamer plies on the lake, making two trips from end to end daily. A number of excellent summer hotels are located here. In the vicinity of Greenwood Lake, —which is often called "Miniature Lake George," from its resemblance in some of its features to that queen of beautiful lakes —are a number of other little less lovely lakes and lakelets. Greenwood Lake may be reached by stages from Monroe, a station on the Erie Railway, fifty miles from New York, or from Greycourt, on the same railway, four miles further on. Another route is via Montclair and Greenwood Lake Railroad. The stage-ride from Monroe and Greycourt is interesting and delightful.

Greenwood Lake, New York.

Devil's Lake, Wisconsin.

AMONG the myriads of lakes glistening among beautiful surroundings in the "Great Northwest" there is none that can be called more wonderful and romantic than Devil's Lake, located thirty-six miles northwest of Madison, not far from the station Baraboo, on the Chi-

Devil's Lake, Wisconsin.

cago and Northwestern Railroad. This charming and mysterious sheet of water, without visible inlet or outlet, is supposed to occupy the crater of an extinct volcano. The gloomy bluffs which wall in the clear cold waters of the lake, and some of which rise over seven hundred feet from its margin, are in striking contrast with the surrounding scenery. The barriers to the lake, beautiful in its strange captivity, are composed of a mass of loosened rocks, piled in grotesque confusion as if hurled aloft in the grim sport of some Titanic race, while the surrounding country is, for a good part, a sandy waste. To the Indian, ignorant of the processes of nature, Devil's Lake was a sacred body of water. Says a writer in a recent number of "Harper's Magazine," visiting at this resort: "The ponderous blocks of Devil's Doorway could only have been placed upon their piers of smaller stones by some superhuman agency, and a cleft rock, perched on a dizzy height, and supported by a single prop of nicely-fitted blocks, was the unquestioned work of some Manitou. Add to this the lake without overflow or source, unprecedented in savage observation, and the effect was overwhelming; and the swarthy hunter, pursuing his game over smiling prairies, came with awe before these strange deep waters in the stern and desolate temple of some unknown deity. The wounded stag, dashing into its cooling waves, escaped pursuit; the very fish roamed in shoals unsought, and so strong was this superstitious dread that the dying warrior perished in agony rather than profane its waters with human lips." As this mysterious and picturesque lake is becoming more widely known it is visited by increasing numbers of tourists, and is invariably included in the round made by "doers" of "The Dells" and other resorts of Wisconsin and Minnesota.

Lake Minnetonka, Minnesota.

WITHIN the borders of a vast forest of hardwood timber extending across the State of Minnesota from the vicinity of Sauk Rapids, on the Upper Mississippi, to near the Iowa line, are hundreds of beautiful lakes of crystal water, which are attractive to the huntsman and tourist. The most fashionable resorts among these lakes is Minnetonka. The name of the

Lake Minnetonka, Minnesota.

lake in the language of the Sioux Indians, who less than a quarter of a century since abandoned its shores, signifies "Big Water." Minnetonka is made up of a series of bays, some twenty-five in number, which form a chain of what appears to be a succession of lakes, which are joined by estuaries, many of which are navigable by steamers. This series of irregular-shaped bays covering an area of over six hundred acres, give ample room for all kinds of rural enjoyment. The heavily-timbered banks, the numerous jutting points and crooked beaches, the stretches of marsh resembling vast lawns, and the numerous picturesque islands, combine to form pictures of varied beauty most pleasing to the eye. Says a recent visitor to Lake Minnetonka: "The Big Woods nearly incloses the lake in its midst, and cosy summer resorts nestle beneath the branches of the great monarchs of the forest, on the banks of the beautiful bays, while villages and hotels have sprung up at the most convenient and available points. Steamers ply on its crystal waters to carry pleasure-seekers to their destination, and fleets of sail and row-boats are to be found at all points of the lake, to supply the demand of fishing parties. Even the least frequented bays begin to show signs of civilization in newly-erected cabins, where some straggling sportsman settles down for a comfortable summer in the deep recesses of the wildwood, where he can be free from the annoyance of the fashionable crowds who frequent other parts of the lake." The lake is located fifteen miles southwest of Minneapolis, and twenty-five miles from St. Paul. Minnetonka Park is the principal resort on the lake. It is reached by the Chicago, Milwaukee and St. Paul Railroad.

Devil's Lake, Dakota.

THIS magnificent sheet of water, called by the Indians "Mini-wakan," is located midway between the Red River of the North and the Missouri River, about fifty-five miles south of the international boundary line. . It is included in an immense body of land which was until very recently supposed to be the property, by treaty, of the Turtle Mountain Chippewa Indians. The present Secretary of the Interior discovered that they had not a shadow of claim to the land, which is now being surveyed and opened to settlement. The water of the lake is dense and salt and green, closely resembling the ocean. The lake is nearly sixty miles in length and fifteen miles in width. It is surrounded by swells of hills varying in height from twenty to two hundred and fifty feet. The northern side is bordered by hills that are well wooded and furrowed by ravines and coulees. The opposite side has less timber, but is also quite hilly. Fish of large size are found in the lake. There are many jutting promontories along the coast, and many handsome islands scattered over its surface, giving a most beautiful effect to the scene which a sail over its waters presents. The land around Devil's Lake is rolling prairie of the richest description. Analysis shows that the water of the lake contains sulphates of soda and magnesia (epsom and glauber salts), and chlorides of soda (common salt), and magnesia. These medicinal properties, taken in connection with the magnificent climate, is certain to attract tourists from all parts of the world, and Mini-wakan will become as famous as a summer resort as the country around it will be for productiveness. A projected branch of the St. Paul, Minneapolis and Manitoba Railway will soon be constructed to Devil's Lake.

The Great Rivers.

And see the rivers how they run
Through woods and meads, in shade and sun,
Sometimes swift, sometimes slow,
Wave succeeding wave, they go
A various journey to the deep,
Like human life, to endless sleep!

PROMINENT among the rivers of the world, for scenery and beauty, rank our own glorious Hudson, the wonderful St. Lawrence, and the picturesque Upper Mississippi. Even the world-famed Rhine is surpassed by the Hudson, which has a considerable advantage in size, though in the length of its navigable portion the latter is a small river when compared with either the Mississippi, the Missouri, the St. Lawrence, the Rhine, or the Danube. The fame of the Hudson—its location and legendry—is known to all the world. The Highlands, incomparable for the combined beauty and majesty of their scenery, and the curious rocky wall, known as the Palisades, extending for miles, from Fort Lee to Piermont, give this noble American river a character wholly its own. Whether viewed from the window of a Wagner car, or from the deck of one of the "floating palaces" borne upon its waters—in the changing light of day, or the mysterious charm of a full moon, the experience of a first glance at its panoramic loveliness will always be remembered. Starting out from the New York wharf for a sail up the Hudson, the scene at the very outset is one of unequalled animation. The river here has broadened into a bay several miles wide, which is covered with craft of every kind—great steamers from over the sea, enormous sailing vessels, crowded ferry-boats, noisy tug-boats, yachts, barges, and fishing boats, all hurrying to and fro in the line of their various missions. Before the limits of the metropolis are passed the scene is changed, and the eye is charmed by the green wooded hills of Westchester on one hand, and the frowning precipices of the Pali-

The Highlands of the Hudson.

sades on the other. For twenty miles this mighty dyke of basaltic trap-rock shuts off the western sky, then suddenly disappears, and the view opens upon the rolling hills and blue outlines of the distant mountains. Then for a score of miles above, the river winds among the rugged mountains of the Highlands, its channel contracted to barely half a mile in width, until at the northern limit of these crags another portal opens and presents to view the beautiful landscape beyond.

Tourists to the Catskills usually include the scene on the Hudson by making the trip up the river by boat. All along the route from New York to Poughkeepsie may be seen elegant summer residences and villas of wealthy New Yorkers, and at various points there are excellent hotels largely patronized during the warm months by residents of the city.

The Upper Mississippi is one of the watercourses much enjoyed by tourists in the summer season. The scenery from Des Moines to St. Paul is varied and interesting, and there are many points of special attraction. A tour through this section gives an opportunity to see the falls of Minnehaha, immortalized by Longfellow, and the famous Dells of the Wisconsin River, second only to the Thousand Islands of the St. Lawrence. These Dells are among the later wonders of our western world. Just before reaching the locality a quick succession of dissolving views is caught through waving boughs of a winding river deep down between massive walls of rock, its silvery surface set with rocky islands capped with green, and the whole crowned by a glorious confusion of receding hills and slopes. The upper and lower Dells form together an irregular gorge some ten miles in length, walled in with sandstone rock

The Dells of the Wisconsin by Moonlight.

from thirty to one hundred feet in height, upon which nature's resource of various design has well-nigh been exhausted. An explanation of this strange and fantastic formation has been given in rhyme :

> " How were all those wondrous objects formed among the pond'rous rocks ?
> Some primeval grand upheaval shook the land with frequent shocks;
> Caverns yawned and fissures widened; tempests strident filled the air,
> Madly urging foaming surges through the gorges opened there ;
> With free motion, toward the ocean rolling in impetuous course,
> Rushing, tumbling, crushing crumbling rocks with their resistless force ;
> And the roaring waters, pouring on in ever broadening swells,
> Eddying, twirling, seething, whirling, formed the wild Wisconsin Dells."

But of all American rivers the St. Lawrence possesses the greatest attractions for tourists. There is not another tour of equal distance in the world that presents such a combination of beauty, excitement, and interest as that across Lake Ontario into the St. Lawrence through the picturesque Thousand Islands and down the wild and boisterous rapids to the metropolis of Canada and the quaint, historic city of Quebec. This trip is generally made by tourists from Niagara Falls, in one direction, or omitting the lake portion, from Kingston on the Grand Trunk Railway, or from Trenton Falls via Clayton, a terminus of the Utica and Black River Railway ; or still again from Cape Vincent, on the New York Central Railway. Almost immediately after setting out from the latter points the steamer enters that portion of the river known as the Thousand Islands. Here, according to the Treaty of Ghent, sixteen hundred and ninety-two islands of various sizes and shapes push their heads up through the waters, and through and among them the river winds its tortuous course. Such a scene as here presents is not to be found anywhere else in the known world. It is a wilderness of islands, some so small as to be barely visible, others acres in extent ; some presenting to the view nothing but bare masses of rock, while others are covered with a thick forest of foliage, green and fresh in summer and tinged with all the colors of the rainbow in autumn. Mighty river and inland sea, mountain and plain, island and continent, nature in her sweet and placid aspect and in her dark and awful mood, all are blended here to form that singular combination of elements which the Iroquois Indians so appropriately named Man-a-to-ana—the Garden of the Great Father. A writer in "Harper's Magazine" thus discourses of the scene : "Islands to right of us, islands to left of us, islands in front of us, lift up their heads, crowned here with jutting rocks, there with forest trees, and again flanked by grassy slopes extending to the water's edge, and fringed with trees whose drooping branches reach down their leafy tips to drink the clear green waters of the river. The view grows more charming as we proceed. Channels open between the islands in every direction, and as our little steamer drives swiftly along the main and broadest channel, the shifting scenes go by us like a panorama. To our left still lies Wells's Island, nine miles long, shutting out all view beyond, while off to the right we catch through the rock-bound channels an occasional glimpse of the American mainland. A run of half an hour more brings us to Alexandria Bay. This is the central point of interest. For ten miles up and an equal distance down the river the islands lie thickest, the cottages are most numerous, and the fishing most alluring. The village, which takes its name from the bay, is perched upon a rocky headland on the American shore." Westminster Park and Poplar Bay are two noted points in this neighborhood, the latter taking its name from a group of five Poplar-trees on the edge of Wells's Island. It looks out upon a great sheet of water, three miles wide and several miles long, studded with islands, whose craggy sides are gray with lichen, spangled with mossy cushions, and belted across with long seams, out of which grow ferns and wild flowers that none can ever hope to touch with human fingers.

Many tourists rush through the Thousand Islands by daylight, in true American style, on a big steamer, drop the morning paper or latest novel just long enough to glance over the rail at a pretty vista of channel or a cosy island home, and imagine they have seen the Thousand Islands. Just so the swift Yankee spends fifteen mortal minutes by the watch in "doing" the Louvre, or St. Peter's, or the galleries at Munich. Whoever does that loses one of the most inspiring opportunities of a lifetime. There is only one such archipelago in the world, and no man looking for the gems of nature's handiwork can afford to sail through the Thousand Islands and not know what they are. To really know what the Thousand Islands are, one should stop among them for at least a week or two, put up at a good hotel, secure a skiff for the term of his stay, and then paddle in and out of these beautiful coves and bays, across and

General View of the Thousand Islands

through these winding and rock-bound channels, and visit island, and promontory, and cliff. He must float slowly over this clearest of all water on a calm day, and see the vast aquarium beneath his keel, where six, eight, twelve feet down through the green sparkling river, is such an under-water garden as the wildest fancy never dared to picture. One of the great attractions of the Thousand Islands is the fishing. It is the great fishing-ground of America. The great catch is pickerel, which may be taken even by inexperienced fishermen, and muskalonge, weighing twenty pounds or less—generally less—are caught in great numbers. The air of the Thousand Islands is heavily charged with ozone, the first effect of which is to induce a delicious drowsiness. The wholesome effect of this air upon consumptives, however, is due not only to the ozone, but also to the piny breezes blowing across the vast Canadian forests, and gathering new richness from the woods of the islands themselves. The island air is, moreover, remarkable for its dryness. The ladies may play croquet in slippers in the early morning without gathering any dampness from the grass; and neither piazzas nor hammocks threaten their occupants even at night with rheumatism or ague. Excellent bathing can be had at many points where sandy beaches are found. In the season there is good duck-shooting upon the river, the birds being mostly of the teal variety. Another water-fowl, passing here under the name of loon, but probably misnamed, frequents these waters in the fall of the year, and stories are told of the immense quantities a skilful sportsman may bag, which need to be taken, as the birds are, *cum grano salis.*

Running the Rapids of the St. Lawrence River.

Leaving the Thousand Islands, the steamer passes almost immediately into the wonderful rapids, and the tourist enters upon the most exciting and exhilarating portion of this trip. The increasing speed of the vessel soon after passing Morrisburg is the signal which sets passengers upon the *qui vive.* The first rapid, or series of rapids, is known as the Long Sault. This is a continuous rapid for nine miles. The river is divided in the centre by an island. In former years the descent of this rapid was made through the south channel only, the north channel being considered too dangerous, but recent examinations have proved that either channel can be descended with safety. The south channel is very narrow, and the swiftness of the current is so great that a raft will drift nine miles in forty minutes, which is equal to the speed of the swiftest steamboats in still water. The rapids of the Long Sault rush along at a speed of twenty miles per hour. The sensation while in this rapid is unlike that when descending its successors. The Long Sault reminds one of the ocean in a storm, except that the swift going down hill in a steamboat is, to most persons, an entirely new experience, and the steep descent is fully realized if one has neglected to take hold of some stationary portion of the steamer. The terrific roar and seething violence of the river is intensely fascinating. Great nerve and power are required in piloting the steamer, so as to keep her straight ahead and in the channel, as a slight deviation would turn the steamer sideways, in which case she would be instantly capsized and submerged. But the discipline and system are so perfect in the management of the steamboat lines that such a thing is never likely to happen. While descending the rapids a tiller is attached to the rudder as an extra precaution, and the force required to keep the steamer straight in her course is so great that four men are kept constantly at the wheel and two at the tiller. Leaving the Long Sault Rapids, and passing through Lake St. Francis, a distance of forty miles, Cedar Rapids are next reached, then the Split Rock Rapids, Lake St. Louis, and the Lachine Rapids. The passage through the Cedars is very exciting. There is a peculiar motion of the steamer, which in descending seems like settling down as she glides from one ledge of rock to another.

A Glimpse of Ottawa.

This is supposed to be owing to the existence of a strong under-current. It was in these rapids that a detachment of three hundred men, under General Amherst, were lost in 1759. The quiet passage of twelve miles through Lake St. Louis serves to stimulate curiosity in regard to the Lachine Rapids, which are nine miles from Montreal, and are the last rapids of importance on the St. Lawrence. The velocity and fierceness of the current are so great, that to avoid the rapids the Lachine Canal was constructed, and, during stormy weather, is used for passage from Lachine to Montreal. The Lachine Rapids are the most difficult of navigation of any on the St. Lawrence. Baptiste, an Indian pilot, has made it his business for over forty years to pilot steamers down these rapids; during the summer season he is exclusively in the service of the passenger steamers, shooting these rapids, and under his skil-

A Glimpse of Quebec.

ful guidance there is entire safety in passing through them. But if the day is stormy, or a south wind prevails, the tourist leaves the rapids behind him with a grateful sense of relief, especially if his point of observation has been the bow of the boat. With rocks ahead and rocks beneath, asserting their presence by impudent thumps against the steamer's keel, the experience is seasoned with just enough thought of danger to give it zest ; and when one is as-sured beyond doubt that there is not the least real danger, the excitement becomes a pleasure.

The trip down the St. Lawrence is incomplete without a visit to the French Canadian cities of Ottawa, Montreal, and Quebec. The scenery in and around all these cities is peculiar and

Cape Eternity and Cape Trinity, Saguenay River.

impressive. Mont real, the metropolis of British North America, from its many commanding features of interest, is the objective for the majority of tour-ists to this section. The city is situated on an island of the same name, lies at the base of Mt. Roy-al, from which the name was taken. The drive around the mountain is delight-ful. The summit is reached by a splen-did carriage-road covering a distance of eight miles, thus rendering the ascent very easy, and from several places during the ride a bird's eye view of the entire city and the ma-jestic St. Lawrence may be had, with the Lachine Rapids in the distance. This mountain, possessing many wonderful natural advantages, is being converted into a magnificent park, which, when completed, will not be excelled in size and beauty. Quebec, founded in 1608, is one of the oldest cities in North America, and also one of the most in-teresting. The plan of the city is nearly a trian-gle, the Plains of Abraham forming the base and the rivers St. Lawrence and St. Charles the sides. The city is divided into two parts, known as the upper town and the lower town. The upper town is strongly fortified, and includes within its limits the cita-

Trinity Cove, Saguenay River.

del of Cape Diamond, which covers the entire summit of the promontory, and embraces an area of more than forty acres. St. John and St. Louis, suburbs, are also included in the upper town. The citadel occupies a commanding site, three hundred and forty-five feet above the river, and is the strongest fortress in America. Quebec is pre-eminently the stronghold of Canada, and is called the "key of the province." The citadel, from its great elevation, affords a fine view of the river and surrounding country. The line of fortifications inclosing the citadel and upper town is nearly three miles in length.

Quebec retains many of the characteristics of its early French founders, and impresses the visitor with the quaintness and venerable air of much that is to be seen, and is suggestive of a little bit of the Old World transplanted to the New. The age of the city shows itself in a marked degree, and the visitor voluntarily accords it a proper amount of respect as an honored relic of by gone days. It has played a most important part in the history of this country, and there is scarce a span's space in the city or its vicinage but what is memorable in history as the scene of some struggle or decisive engagement. The French language is the exclusive medium of intercourse among many of the inhabitants. The drives around Quebec are full of interest and afford delightful prospects. Eight miles below the city are the celebrated Falls of Montmorenci. As is well known, these falls are only fifty feet wide, but descend in a perpendicular sheet more than two hundred and fifty feet. The place is much frequented.

A very pleasant rounding off of the St. Lawrence tour is made by including a trip to the remarkable Saguenay River, its largest tributary. Leaving Quebec, a detour of two days affords the opportunity for viewing the grandest and most striking river scenery on this continent. At Tadousac, 120 miles below Quebec, the Saguenay empties into the St. Lawrence, and from the moment the channel is entered the beholder is impressed with the grandeur of the prospect before him. On either side perpendicular cliffs of granite and syenite in solemn majesty rise abruptly from the water's edge to a height of nearly 2000 feet. The quiet flow of the river in its deep and rock-bound channel is in perfect accord with the wondrous charm of the situation. The depth of this river is something remarkable; at its mouth a line of 330 fathoms could not sound bottom; at St. John's Bay, 28 miles above Tadousac, the water is one mile and a half deep. Six miles beyond St. John's Bay is Eternity Bay. Two majestic promontories, like gigantic sentinels, guard its entrance; Cape Trinity, 1500 feet high, on the left; Cape Eternity, 1900 feet high, on the right. At this point the river is a mile and a quarter deep. The headwater of the Saguenay is the Lake St. John, 40 miles long and nearly as wide, and although eleven rivers flow into it, its only outlet is the Saguenay. The original name of the latter was Chicontini, an Indian word, signifying Deep Water. Sixty miles above Tadousac is Grand or Ha-ha Bay, nine miles long and six wide. It affords good anchorage for the largest vessels, the average depth being from 15 to 35 fathoms. The attractions of this place are many and very inviting. Its name is said to come from the joy it afforded the first navigators of the river, who found here their first landing-place, and expressed their delight by a hearty Ha! ha! The Grand Trunk Railway, and the Royal Mail and Richelieu Line of steamers comprise the favorite lines of travel to and from all points in Canada and the St. Lawrence River.

Trenton Falls, New York.

In all the world there is, perhaps, no stream which in the same space presents so many and various shapes of running and falling water, as does the West Canada Creek at Trenton Falls. This lovely spot, to which thousands now annually resort, embraces scenery altogether unique in its character, as combining at once the beautiful, the romantic, and the magnificent—all that variety of rocky chasms, cataracts, cascades, rapids, etc., elsewhere separately exhibited in different regions. Trenton Falls is situated in the central part of New York State, on the line of the Utica and Black River Railroad, eighteen miles north of Utica. The Falls, consisting of five cataracts and a series of cascades of unexcelled picturesqueness and beauty, are a part of the West Canada Creek, the main branch of the Mohawk River, as the Missouri is of the Mississippi, having lost its proper name because not so early explored. The Indians gave to the Falls the beautiful and descriptive name of "Kang-a-hoo-ra"—Leaping Water—and such it literally is, for here the stream flows for two miles through a ravine or chasm from seventy to two hundred feet deep, descending three hundred and twelve feet in one continuous succession of cataracts, cascades, and rapids—rushing, roaring, dashing, whirling, leaping and plunging throughout the entire course. A tour of the ravine is made by paths cut in the sides of the rocks, and by staircases leading from the lower to the higher points, as at Watkins' Glen and Ausable Chasm. By the changes made in these paths during the past two years new views have been opened from the heights, several of which present scenes that neither pen nor pencil can catch. The descent into the ravine is made easy and safe by five pairs of stairs with railings, leading from the bank down one hundred feet to a broad pavement, level with the water's edge, a furious rapid being in front that has cut down the rock still deeper.

Rocky Heart, Trenton Falls.

This is at the lower end of the gorge. The first impression of the tourist upon reaching this subterranean world is astonishment at the change; but, recovering instantly, his attention is forthwith directed to the magnificence, the grandeur, the beauty, and the sublimity of the scene. The paths extend for upward of two miles along the water's side, introducing the visitor at every step to indescribably beautiful and ever-changing views. The principal falls are the Lower Falls, thirty-three feet in perpendicular height; Sherman Falls, thirty-five feet, from a shelving rock into a dark pool below; High Falls, one hundred and nine feet; and Mill-dam Falls, fourteen feet in height. In a story called "Edith Linsey," written by the late N. P. Willis, occurs a description of Trenton Falls, of which the following is an extract: "Most people talk of the sublimity of Trenton, but I have haunted it by the week together for its mere loveliness. The river in the heart of that fearful chasm is the most varied and beautiful assemblage of the thousand forms of running water that I know of in the world. The soil and the deep-striking roots of the forest terminate far above you, looking like a black rim on the inclosing precipices; the bed of the river and its sky-sustaining walls are of solid rock, and, with the tremendous descent of the stream—forming for miles one succession of falls and rapids—the channel is worn into curves and cavities which throw the clear waters into forms of inconceivable brilliancy and variety. It is a sort of half twilight below, with here and there a long beam of sunshine reaching down to kiss the lip of an eddy or form a rainbow over a fall, and the reverberating and changing echoes,

'Like a ring of bells, whose sound the wind still alters,'

maintain a constant and most soothing music, varying at every step with the varying phase of the current.''   In one of his letters the same writer says: '' Trenton Falls is the place above all others where it is a luxury to stay—which one oftenest revisits—which one most commends to strangers to be sure to see.''

The Seashore

"Thou glorious sea! More pleasing far
When all thy waters are at rest,
And noonday sun or midnight star
Is shining on thy waveless breast.
Yet is the very tempest dear,
Whose mighty voice but tells of thee;
For wild or calm, or far or near,
I love thee still, thou glorious sea!"

FROM the pine woods to the Everglades the historic Atlantic washes the shores of our broad land, now "dashed high on a stern and rock-bound coast," and eddying around the northern islands; now flooding through the Narrows to bathe the feet of the metropolis, and ebbing back by Coney Island and Sandy Hook; anon sweeping with long fine swells into the Sounds of Albemarle and Pamlico, and storming by Cape Hatteras, then capping against the coral reefs of the Florida Keys, and meeting the warm waters of the Gulf; while across the continent, three thousand miles away, the great Pacific's blue waters sparkle through the vista of the Golden Gate, and chant the vespers for the quaint and ancient white-walled Missions amidst the vineyards of Southern California. Thus, with a "deep and dark blue ocean" rolling on either side, the American is most naturally a lover of the sea, and a frequent pilgrim to its shores. It is probably true that a majority of those who annually give themselves a season of relaxation seek it along the many curves of glistening sand, where the waves of

ocean beat more gently. So general has this seaward tendency become that the "rapture on on the lonely shore" is often intruded upon.

Along the barren sands cities have sprung up; the trackless wastes are paved by corporations; the moonlight that sends a silver path across the sea pales on the land before the electric light that glitters in the gay pavilions and hotels. Where once the only music that trembled through the air were the deep chords from the booming billows, the sounds of merry songs and the notes of tinkling cymbals and stringed instruments now fill the air; and the fresh breeze, which once had far to wander to blow its good to humankind, now fans the fair brow of beauty, driving on the sea-road or promenading on the ocean-walk, while the forgotten untutored mermaids hang their heads for shame beneath the waves. No American resorts are so cosmopolitan. All countries, all tongues, all conditions are represented among those who flock thither for amusement, for health, for the love of excitement, for the love of the grand ocean. They ramble on the shore, they sit on the beach, they sail through the waters, they gather the shells, they sport in the surge. Though some may have to search for that solitude so dear on occasions, there is room for all, and one may always find some nook or corner where he may follow his own bent of worship. To the invalid and weary the salt breeze comes like a cool hand on the fevered brow, and bears the healing of the seas; to the strong worker what a blissful rest to cast himself upon the sands, forget the dusty town behind in the contemplation of these vast waters that bear the white-winged messengers of commerce and beat about a thousand lands. In the words of Oliver Wendell Holmes: "Who does not love to shuffle off time and its concerns at intervals,—to forget who is President and who is Governor, what race he belongs to, what language he speaks, which golden-headed nail of the firmament his particular planetary system is hung upon, and listen to the great liquid metronome as it beats its solemn measure, steadily swinging when the solo or duet of human life began, and to swing just as steadily after the human chorus has died out and man is a fossil on its shore." To Byron the sea was power, solemnity, eternity,—the "glass upon which the face of the Almighty is seen." To the poet Procter it was:

> "The sea! the sea! the open sea!
> The blue, the fresh, the ever free!
> Without a mark, without a bound
> It runneth the earth's wide region round;
> It plays with the clouds, it mocks the skies,
> Or like a cradled creature lies."

But romance aside, the great desideratum of a vacation is to most of us recuperation in health and strength; and the question must be determined in each case individually, whether the sea or mountain air is most beneficial. The distinctive feature of the seashore is the opportunity it affords for salt-water or surf bathing. The great majority of persons are more or less benefited by sea-baths, when taken under proper conditions, and with the observance of proper precautions. But it is generally admitted among bathers of experience that the effect of these baths may be either stimulating or depressing—that they may do great good or much harm, according as all the conditions may be understood and observed. The sea is a powerful chemical agent. Many of the salts held in solution in its waters possess strong medicinal properties, which act directly through the pores of the skin. In all cases where there is reason to question the expediency of a course of sea-baths, or to suspect a tendency to heart disease, medical advice should be taken and carefully followed. Few general rules can be laid down, but it may be set forth as one of them that a short time in the water is always best; and that the bather should hasten briskly to the dressing-room after leaving the water, and indulge a vigorous rubbing with coarse towels.

Atlantic City, N. J.

THIS "City of Homes," located on the Atlantic coast, sixty miles southeast of Philadelphia, is one of the largest and most popular watering-places on the Atlantic seacoast. During the season tourists in great numbers from every quarter are drawn to it; but being very convenient of access to Philadelphia, the greater proportion of its visitors and summer residents are from that city. Atlantic City is really situated on an island formed by the Atlantic Ocean on the east, and a navigable strait—The Thoroughfare—on the west, Absecom Inlet on the north, and Old Inlet on the south.   It contains a large number of private cottages of exceeding beauty, both in themselves and their surroundings.   The peculiar dry atmosphere of this resort, its magnificent beach—one of the best and safest on the coast—its splendid bathing and comfortable hotels and cottages, have made it one of the most popular watering-places on the coast. It is not only a city in name, but in fact, possessing all the conveniences enjoyed by cities of a larger growth.   The resident population is six thousand, and those who visit there in search of health or pleasure will find no lack of necessities and all of the luxuries enjoyed at home. There are fine markets, good stores in great variety, street cars, stage lines, unexcelled livery, gas, first-class medical attendance, and stores, etc., etc.   There are pleasant drives for many miles up and down the beach, which · is smooth, sandy and gently sloping.   For two miles along the ocean front of the city the beach is skirted by a broad plank walk, which is the favorite resort for promenading during bathing hour and evening.   The surrounding country is entirely destitute of attractiveness to the vision, consisting as it does, for the most part, of wide-stretching salt marshes ; but because of the fish that may be taken in its waters, and the game that may be bagged from among the reeds and rushes it is a sort of Paradise to the persistent sportsman.   The run from Philadelphia to Atlantic City by the West Jersey Railroad, controlled by the Pennsylvania, is made in ninety minutes, and passenger trains are run often enough every day to suit all desires and necessities—in fact so fast and frequent are they that Atlantic City has come to be regarded as almost a suburb of the Quaker City.

Cape May, N. J.

FEW seaside resorts in the country surpass Cape May in the points of attraction chiefly considered in estimating the merits of such a resort.   Good, safe bathing, pleasant drives, and desirable society are the first considerations of a sojourn at the seashore, because there are few resources aside from these.   A "Summer City by the Sea" is often in its appearance a rude shock to preconceived ideas.   Every circumstance, however, attendant upon a first introduction to Cape May is calculated to satisfy all expectations.   A swift two hours' ride over the West Jersey Railroad from Philadelphia precedes arrival at the handsome depot of the Company, upon the very sands of the ocean.   The traveller glances at the curling crests of white foam, accompanied with long-drawn inhalations of the salt sea air, and is whirled through the little village to his chosen hotel.   It is a peculiarly situated place, where the southern boundary of New Jersey becomes drawn out into a long pointed strip of land, and projects far down into the ocean.   It would seem as if the forces which had been at work in ages past chiselling out bays and headlands had foreseen a coming need, and, working with intelligent purpose, had left this strip of land that, upon its very point, where the waves welcome the noble Delaware to their embrace, might be situated a famous watering-place.

Cape May, to quote a well-known writer, "possesses one of the few really fine ocean beaches of the world; a splendid expanse of smooth white sand, firm yet soft to the tread, stretching for miles up and down the coast, over whose slight incline the waves break with a regularity and gentle force which makes life-lines entirely unnecessary.   The hotels and cottages are in close proximity to the beach, which circumstance, taken in connection with the fact that there are no unsightly stretches of barren land or wastes of salt marshes to offend the eye, forms a prominent factor in the exceptional popularity of the place.   The situation of the Cape gives it peculiar and decided advantages as a sanitarium.   It is surrounded on three sides by the Atlantic, whose purifying breezes fan it without stint, and afford almost entire immunity from that pest of sea-side resorts, the mosquito.   It was this natural adáptation to the purposes of both health and pleasure that made Cape May a favorite resort long before the era of railroads. It was sought by thousands in that early time who loved the sea for its beauty alone, and it will continue to preserve its prestige unbroken so long as man continues to be endowed with the capacity to enjoy. The patronage of the place has long been monopolized by a class which represents at home the highest grade of intelligence, social standing, and culture. Ten thousand annual summer visitors from Baltimore, Philadelphia, Washington, and Pittsburgh, representing the best families of those cities, have left the impress of their character upon the town, encouraging that thrift and enterprise and good taste in its citizens which builds up beautiful avenues, preserves a cleanly, wholesome condition of its streets, and creates a strong, healthful public opinion in the direction of morality, culture, and refinement.''

Cape May, from the Pier–Stockton House.

A magnificent drive, fifty feet wide, extends along the whole sea front, flanked on the ocean side by a board walk ten feet wide. These are constructed in the best manner, the drive being well gravelled, and connecting as it does with the principal streets of the town, forms a continuous circuit of many miles, combining the unsurpassed ocean scene, and the most attractive views of the city. The board-walk sweeps along in graceful curves for a distance of near two miles, and as smooth as a ball-room floor, commanding an unobstructed prospect of the bathing-grounds on the one side, and the carriage-way on the other. The principal avenues of the city are covered with shells from the sea, thus rendering them free from dust, and delightful for promenaders and others visiting the handsome shops, hotels, and private residences extending along them. The favorite hotel at Cape May, the Stockton, is under the management this season of Col. J. F. Cake, well known as the proprietor of Willard's and the Metropolitan, of Washington, in years past, and in his hands this palace by the sea is likely to become more popular than ever.

Long Branch, N. J.

AMONG the oldest as well as the most fashionable and popular resorts on the New Jersey coast is Long Branch. It has an unusually fine beach for bathing and promenading, and possesses in perfection the best attractions sought on the seashore. A new feature in recent years is the great iron pier extending from the bluff out some 800 feet into the ocean. "The bluff" is a sandy elevation, rising abruptly from the beach to a height of twenty feet, forming a plateau upon which the hotels and residences are located, overlooking the boundless expanse of ocean. It extends, in an almost unbroken line, five miles. The iron pier, with its top on a level with this bluff, reaches far out beyond the breakers, and furnishes a long prom enade, as well as a convenient fishing stand. There is a restaurant on it, also accommodations for an orchestra, while underneath are numerous bath-houses.

The drives about Long Branch are a feature, and the famous "Beach Drive," extending a distance of twenty miles or more, commands a fine view of the sea for almost the entire distance. From some of the hotels may be seen showy equipages in passing and repassing lines, pleasantly breaking the vision of the bright green of the lawn, and the deep blue of the ocean beyond. There are at Long Branch no salt marshes, sandy plains, nor mosquitoes. The soil from the bluff back is of the most fertile character, and the art of man working upon this and aided by unlimited capital, has done so much to beautify the place and add to the great natural attraction of the sea, that it can never be in any danger of losing its high rank and prestige. Owing to the fame of Long Branch as a resort of the fashionable classes, there exists in the minds of a majority of those who have never visited it an impression that it is unsuited to people of moderate means and quiet tastes. This is, however, a great misconception. The charges at the hotels range from three dollars to four dollars per day, and from twelve dollars per week upwards. The carriage fares are also very moderate, and need exclude none who do not own their own teams—and these constitute three-fourths—from the pleasures of the drive. There is less attempt at vain display, and less excitement than at many less noted and cheaper resorts. Fashion decrees no particular course of conduct, or style of dress, and there is enough democratic leaven in the lump to make it proper for every one to do as he pleases, provided the ordinary proprieties of life are observed. There are several ways of reaching Long Branch from New York, viz.: The Pennsylvania route, from Desbrosses and Cortlandt streets, *via* Newark, Rahway, and Amboy; or by steamer, running four times daily in summer from Pier 14 to Sandy Hook, thence by rail; or the all water-route by steamer from foot of Twenty-second street; and the all-rail route, *via* Long Branch Division of the New Jersey Central Railroad.

Ocean Grove and Asbury Park, N. J.

A LITTLE over ten years ago several ministers and laymen of the Methodist Church, having in view the establishment of a camp-meeting ground, purchased the tract of land upon which the beautiful little city of Ocean Grove is built, at a cost of six hundred dollars. The splendid beach, fine location, and beautiful surroundings were expected to make it popular as a camp-ground, but little did its projectors at the outset dream that in less than a decade it would become a city of elegant, costly and substantial cottages, and one of the famous and prominent seaside resorts of the Atlantic coast. The advantages which the place offered for a summer home by the ocean were soon discovered, however, and the plans of the founders were enlarged, and the design of establishing here a summer retreat for Christian families was conceived. The plot of ground had been dedicated to religious purposes, and chartered under the name of the "Ocean Grove Camp Ground," and its improvement with a view to establishing a city was now systematically begun. The association is authorized to make its own laws, and they have framed these so as to secure, for all time, the purposes had in view when the work was commenced. No intoxicating drinks are permitted on the ground ; boating, bathing, and driving are strictly prohibited on Sunday ; and all behavior unbecoming the repose of such a place is at once suppressed. These regulations, and the natural advantages of the location, make it a pleasant and quiet place, where families can remain free from intrusion and annoyance, and where the beneficial effects of sea-air and sea-bathing can be enjoyed without the expense and tax upon the system exacted by resorts at which fashion and folly too often rule. In August of every summer camp-meeting is held on the grounds reserved for the purpose, continuing two weeks. Ample provision is made for the immense number of people who visit the Grove during the camp-meeting season under tents, which may be rented at reasonable rates. Asbury Park adjoins Ocean Grove, being separated therefrom only by Wesley Lake, a narrow but beautiful sheet of water. Asbury Park is an offshoot, so to speak, of Ocean Grove, but it is less strict in its police regulations than the latter, and on that account is preferred as a residence by some persons. Like its parent city, it has a fine beach and splendid bathing facilities. Its streets are regularly laid out and adorned with beautiful shade trees. Both Ocean Grove and Asbury Park are provided with excellent hotels and numerous and good boarding-houses. These resorts are located about six miles south of Long Branch, on the New Jersey coast, and may be reached by the Pennsylvania and New Jersey Central railroads.

Newport, R. I.

NEWPORT is called the "Queen of American watering-places." For salubrity of climate and beauty of surrounding scenery it is claimed to possess advantages over all other similar resorts. Whether these claims be well sustained or not, to Newport are gathered every summer the first elements of American society in greater numbers, perhaps, than they are found elsewhere. It is also the favorite summer resort of foreign residents in America ; and several of the ambassadors from Europe have cottages there. These give it a social aspect of the highest charm. The city is adorned with villas of the most costly and ornate character—surrounded by every feature of wealth and rural luxury—the country seats of gentlemen of fortune and culture, of New York, Boston, and other cities. At no other American resort are balls, receptions, and dinner and garden parties given on such a lavish and tasteful scale, and at no other place on our shores can such a perfect whirl of superb equipages be seen as may be beheld every evening on the grand drive on Bellevue avenue, rivalling in number and elegance those of Hyde Park and the Bois de Boulogne. Its site is matchless, its climate delicious, its bay glorious. The grandest boats that steam over the seas of the world land tourists at Newport.

The most beautiful and swiftest flying yachts that skim upon the waters of the globe spread their white sails about the shores at Newport during "the season." In and around the city are many interesting and beautiful localities. Buildings erected long before the Revolutionary War, and occupied during the period of the struggle by Rochambeau and other heroes of distinction, are still standing. Among the scores of other natural and artificial curiosities which contribute to the charm of the place, may be enumerated the "Old Stone Mill," supposed to have been built by the Northmen several hundred years before Columbus discovered America ; Fort St. Louis, a quaint old ruin at the entrance to the harbor ; Fort Adams, one of the largest fortifications in America, situated on a point a mile and a half southwest of the city ; Purgatory Rocks, Hanging Rocks, "The Dumplings," and the Glen, wonderfully strange natural formations in the cliffs along the shore, and in the rocks in the harbor ; Touro Park, given to the town by Judah Touro, a Hebrew, who was born here, and the Jewish Cemetery and Synagogue, preserved through bequests left by him. The visitor can occupy days in studying these and other attractions of the place, and in the pursuit will find information as well as pleasure. The beaches at Newport are exceedingly beautiful, and the bathing is unsurpassed by any seaside resort in America.

Newport is one of the capitals of Rhode Island, and is situated on a declivity of the southwest shore of the island from which the State is named, facing the harbor on Narragansett Bay. Its older portion, lying near the wharves, has many narrow streets, bordered with the residences of the permanent inhabitants, many of which are mansions of "ye olden time." New Newport almost surrounds the old town, and stretches away to the south with a great number of villas and cottages, of which we have before spoken. This resort may be reached from New York by the Sound line of steamers, or by the Short Line Railroad, and from Boston, the Old Colony or Boston and Providence Railroads.

Narragansett Pier, R. I.

THE fame of Narragansett as a summer resort has been wholly achieved during the last twenty years, the first sojourn of boarders in the locality having occurred, according to present traditions, in 1856. Two or three years later visitors began to multiply, until the place developed into one of considerable popularity. It has now an array of hotels almost too numerous to mention, with various capacities from fifty to three hundred guests, and rates from eight dollars per week to three dollars per day. The convenient location of Narragansett Pier at the mouth of Narragansett Bay, and convenient to so many large cities, together with other natural advantages, furnish good reason for its large patronage, and render its popularity reasonably well assured for future years. It has a fine beach both for driving and bathing, which, with the fine fishing and sailing make its advantages marked. The surf is light, and the water deepens very gradually, which, with the absence of strong currents renders it more than ordinarily safe. One great attraction is the delightful boating on the bay. The largest hotel is situated on Narragansett Heights, three miles from the Pier, from which there is a grand view of the ocean and adjoining country. In this vicinity also is Silver Lake, a picturesque and enchanting spot. Narragansett Pier is most directly and pleasantly reached by the elegant boats of the Stonington Line from New York.

Martha's Vineyard, Mass.

FOR over fifty years this island has been a prominent camp-meeting ground for Methodists and Baptists, who have congregated here during the summer for worship in "God's first temples." Of late years it has become very popular as a place for summer residence. The island, which is about twenty-two miles in length, and from six to ten miles in width, was discovered in 1602 by Captain Goswold, who gave it the name it still bears. Martha, it is said, was a "lost Lenore" of Captain Goswold. Oak Bluffs, the principal resort on the Island, has been aptly called the Cottage City of America. It contains over one thousand cottages, most of which are elegant and many of them very costly. Life at Oak Bluffs is peculiarly free from the restraint and care which characterize most fashionable seaside resorts. While the place is still largely frequented by Methodists, sojourners of this denomination by no means monopolize the city. A camp-meeting, usually lasting for two weeks, is still held every summer in the month of August. Some of the most distinguished and eloquent Methodist preachers may be heard during "camp." The fine drives in the vicinity of Oak Bluffs afford many charming ocean views. Boating, sailing, sea bathing, blue-fishing, and other seaside pastimes, charm and delight the visitor throughout the season, which reaches its height during "camp-meeting week," when from 25,000 to 40,000 people are on the ground. Oak Bluffs is well provided with excellent hotels, but it has few boarding-houses. Rooms can be readily rented at private houses and meals obtained at the restaurants and hotels. The island of Martha's Vineyard constitutes a county of Massachusetts and is separated from the mainland by Vineyard Sound. There are other resorts besides Oak Bluffs on the island—resorts of considerable pretensions and of high grade. Two miles west of Oak Bluffs, on the excellent harbor known as Holmes' Hole, is the town of Vineyard Haven. Many summer boarders sojourn here. Edgartown and Katama, seven miles distant from the Bluffs, and connected with it by a narrow gauge steam railway, affords fine facilities for bathing and boating. Boats run daily from New Bedford and other points across Vineyard Sound to the various landings on the island.

Nantucket, Mass.

A QUARTER of a century ago the quaint old town of Nantucket, situated on the northern shore of the island bearing the same name, was a busy and prosperous town, but with the decline of the whale fishery its industry was ruined. Its chief business now is to entertain the army of summer tourists who yearly seek recreation and rest on the island. Nantucket is famous for its blue-fishing, which is indulged in not only by trolling, but by beach fishing, known as heave and haul, that is, casting a line from the shore among the breakers and hauling it in quickly. Riding and sailing are also among the favorite exercises of this resort. Surf-bathing is somewhat unsafe, and hence but little indulged in. The sea air is fresh and invigorating at all times—the sojourner here being practically at sea. The thermometer rarely rises above eighty degrees, and the nights are always deliciously cool. With the healthfulness of its climate, its quiet repose, and general home-like character, it offers strong attractions to the invalid, while it invites all to its recreations and rest from the activities of life. Tourists are invariably charmed and delighted with Nantucket as a resort. On the eastern side of the island, situated on a bluff, is the little village of Siasconset. It is quite a favorite resort, and is annually visited by many tourists. Nantucket is about three hours by steamer from New Bedford.

Bass Rocks.

THIS is a new summer resort on Cape Ann, thirty miles from Boston by the Eastern Railroad. It is situated on the high rocky shore of Gloucester, Mass., between Eastern Point and Rockport, and commands magnificent views of land and ocean. Good Harbor Beach, three-quarters of a mile in length, is the finest on the " North Shore " for surf bathing, and is safe at any time of tide. There is also a shallow inlet, with clean sand bottom, for those who prefer still-water bathing. The drives around the Cape, and within a circuit of twenty miles inland, are all very beautiful. The amusements are dancing, billiards, bowling, sailing, and deep-sea fishing. Bass Rocks is one and a half miles from the railroad station, and is connected by telephone with all telegraphic lines.

From Cape Ann to Cape Cod.

THAT portion of the Massachusetts coast stretching northeast from Boston to Cape Ann, and southeast to Cape Cod, with its thousand strange and beautiful indentations and jagged outlines, is famous among travellers and tourists of our land. Boston, the metropolis of New England, and centre of this section, is noted for its historical places of interest, and the beauty of its suburbs. The Common, that great promenade and favorite play-ground for children, almost completely shaded by noble old elms, with its magnificent soldiers' monument, and beautiful Brewer Fountain ; that pink of petite parks, the Public Garden, with its statues and monuments, lakelets and fountains, shady nooks and flower-bordered paths ; the many historic old buildings, such as Old South Church, desecrated by the British army during the War of the Revolution, Burgoyne having turned it into a cavalry school for his troopers, Faneuil Hall, termed the "Cradle of Liberty," where American patriots first resolved to resist the exactions of the British Crown; the Old State House; Bunker Hill Monument; the Public Library; the art galleries, and a thousand other places of quite equal interest, which might be named, combine to make Boston a favorite city to tourists. Passing along the coast of the harbor toward the northeast the first seaside resort reached is Chelsea or Revere Beach, four miles from the "Hub," and accessible by steam and horse cars. The only attraction here is the fine beach, and the excellent suppers served at its hotels and restaurants. Lynn, the famous "shoe town," eleven miles farther to the east, is a pretty city with beautiful surroundings. Long Beach, stretching from Lynn Bostonward to Nahant, affords one of the finest beach drives in this country. Nahant is a picturesque and attractive resort, combining, perhaps, more varieties of ocean scenery and general pleasure advantages than any other sea-side town on the New England coast. Swampscott, twelve miles from Boston, is next reached. This is one of Boston's most fashionable watering-places. It has four beautiful hard sandy beaches, which afford perfect seaside walks and drives. The surf-bathing here is admirable Four miles farther on is the historic old town of Salem, a very agreeable place of summer residence. Farther on, twenty miles from Boston, is Marblehead, "a backbone of granite, a vertebra of syenite and porphyry, thrust out into Massachusetts Bay in the direction of Cape Ann, and hedged about with rocky islets." Marblehead is one of the most famous of American cities. Its situation is such as to command a beautiful view in all directions. Perfect surf and still-water bathing, excellent fishing, and the general healthfulness of the climate combine to make it a popular resort. Manchester, five miles from Marblehead, is the next point reached. This is one of the loveliest watering-places on the Massachusetts shore, as it is one of the most famous among tourists and travellers of our country. Gloucester, a pleasantly situated and compactly built city, comes next in the line. It is twenty-eight miles from Boston, and while resorted to by many during the summer months, offers fewer attractions to tourists than most of the other resorts along the coast. Fishing is the all-absorbing industry of the place. Rockport, on Cape Ann, thirty-one miles from the "modern Athens," has in recent years gained considerable popularity as a summer resort. Many beautiful cottages, the summer homes of Boston merchant princes, are erected here. Granite quarrying is the aristocratic and money-making occupation of the people of Rockport. Pigeon Cove is situated at the extreme point of Cape Ann, and is by reason of the great beauty and the sublimity of its scenery, the healthfulness of its climate, its medicinal springs of "true chalybeate mineral water, having decided tonic properties," its splendid surf and still-water bathing, a much-frequented and deservedly popular ocean-side resort. Turning back to highly-cultured Boston, and starting down the harbor, bound for Cape Cod, many natural beauties, calculated to delight and interest the tourist, are encountered. It is not easy to have a more delightful sail than down Boston harbor, when its islands and the banks of its shores are clothed in their summer garb. Beautiful villas fringe the southern coast of the bay from Boston to beyond Cohasset. Quincy, eight miles from the Hub, and the birthplace of John Adams, second President of the United States, has become one of the most select seaside resorts in the neighborhood of the New England metropolis. Weymouth, five miles farther down, is noted for fine summer residences, and the excellent facilities it offers for bathing, fishing and sailing. Hingham, fifteen miles from Boston, is celebrated for the beauty of its scenery, and the superiority of all of its seaside features. Melville Garden, Downer Landing, an alluring retreat, hard by, is daily sought by crowds from the city. It is, perhaps, the most popular place in the harbor for "spending the day." Its fine grove, excellent restaurant, commodious dancing pavilion, and superior clam-bakes are widely known among Eastern Massachusetts folk. Nantasket, but a mile farther from Boston, is famous for its unrivalled four-mile beach, its elegant bathing, and good hotels. Cohasset, six miles farther on, affords a good opportunity for seeing all that is grand and sublime in old ocean. The surf-bathing here in calm weather is superior, though during a gale the sea becomes very rough, and the in-rolling waves rise extremely high. Scituate, six miles from Cohasset, is very much like it in character. Plymouth, the American Mecca, would commend itself to tourists from all parts of the world, because of its historic associations, even though it were not, as it really is, one of the most delightful of watering-places on the American coast. Sandwich, Cotuit Port, Yarmouth, Hyannis, and Wellfleet, "on the cape," are all situated in the midst of charming surroundings, and are favorite resorts. Provincetown, on the extreme point of Cape Cod, is becoming quite popular as a summer resort, though by reason of the sterility of the soil in the neighborhood, which has made it undesirable as a place of permanent residence, and a good place to emigrate from rather than go to, it has not received as high a rank among watering-places as it merits. All of the resorts mentioned under this head may be reached by boat or rail from Boston, and at all good hotel accommodations will be found.

Bar Harbor, Mt. Desert Island, Me.

MT. DESERT ISLAND, in Frenchman's Bay, just off the coast of Maine, about one hundred and ten miles east of Portland and forty miles southeast of Bangor, has, by reason of the coolness of its climate and the magnificence of its scenery, become one of the most popular

Bar Harbor and Mt. Desert.

summer resorts for tourists along the New England coast. The island is about one hundred square miles in extent and has a population of four thousand. The greater part of its surface is covered with thirteen granite mountains, whose highest peak, Green Mountain, attains an altitude of some fifteen hundred feet. High up among the mountains are many beautiful lakes, the largest of which is several miles in length. These lakes, and the streams that flow into them, abound in trout. The southeast coast of the island is lined with stupendous cliffs several hundred feet in height. The best description that can be given of the island is the language of Mrs. Browning:

> " An island full of hills and dells,
> All rumpled and uneven,
> With green recesses, sudden swells,
> And odorous valleys, driven
> So deep and straight, that always there
> The wind is cradled to soft air."

Bar Harbor, on the eastern shore of the island, is the favorite stopping-place for tourists. The village here is locally known as East Eden, and contains a number of first-class summer hotels, chief among which is the Grand Central. From Bar Harbor the visitor is afforded the best opportunity to explore the cliffs on the shore and make excursions to points of interest in the interior of the island. Among the objects of interest in the vicinity of this resort are Green Mountain, which may be easily ascended, the scenery from the summit of which is extremely grand and beautiful ; Schooner Head, a mass of white jutting cliffs, which from the sea bear a close resemblance to a ship under sail ; The Ovens, a number of holes resembling in form a " Dutch oven," worn in the rocks by the action of the tides, approachable only when the tide is out ; Great Head, the highest headland between Cape Cod and New Brunswick ; Thunder Cave and Spouting Horn, two mysterious caverns in the rocky wall surrounding the island, and many other strange and charming places. Boating and fishing on the bay and angling in the lakes is the favorite pastime with tourists domiciled here. The most popular way of reaching Bar Harbor is by steamer from Portland.

Isles of Shoals, N. H.

ABOUT nine miles off the coast of New Hampshire, southeast from Portsmouth, is a group of small isles known as the Isles of Shoals, which are popular, particularly among New Englanders, as a summer resort. The history of the Isles dates back to July 15th, 1605, when they were seen by the French navigators, De Monts and Champlain. They were very early the resort of fishermen, and were the home of a large and busy community of traders and fishermen by the middle of the seventeenth century. The people were ordered off the islands at the outbreak of the Revolution, and a few only returned at the close of the war, from which time the population has gradually diminished, until now the islands are simply the temporary abode of the " valetudinarian and the summer idler." Appledore, the principal island of this barren group, rises to a height of about seventy-five feet above the level of the sea. Star Island may, perhaps, be reckoned as second in importance On both of these islands large and commodious summer hotels are located. Other dark and gloomy ledges, which rear their heads above the roaring breakers, and upon which many a stanch ship has been dashed to pieces, are known as Smutty Nose, Londoner's, Duck, and White Islands. Fishing and boating, which are unsurpassed here, contribute to make the time pass agreeably with tourists. The atmosphere is quite bracing at all times, and for that reason it is not advisable for persons afflicted with pulmonary ailings, and others with delicate constitutions, to go directly to the Isles from warm and quiet inland places. Mrs. Celia Thaxter, the authoress, a native of Appledore, has made famous the Isles of Shoals by many glowing descriptions of their charms in her writings. The hotels and boarding-houses are quite liberally patronized throughout the entire season. Steamers ply between the Isles and Portsmouth, ten miles distant.

THE GRAND TRUNK RAILWAY,

IN CONNECTION WITH THE

Richelieu and Ontario Navigation Company,

— IS THE —

GREAT PLEASURE ROUTE,

And they now offer a choice Selection of Popular Excursion Routes,
via Niagara Falls, Toronto, River St. Lawrence (with
its Thousand Islands and Rapids), Montreal,
Quebec, River Saguenay, Lakes Cham-
plain and George, Saratoga, &c.

The Richelieu and Ontario Navigation Company's Steamers comprise the original Royal Mail and Richelieu Lines, with the addition of several new steamers, thus forming two first-class lines of steamers, which, for speed, safety, and comfort, cannot be surpassed. They are the only lines now offering tourists an opportunity to view the magnificent scenery of the Thousand Islands and Rapids of the St. Lawrence, also the far-famed River Sague-nay. This route possesses peculiar advantages over any other, as by it parties have their choice of either side of Lake Ontario, and River St. Lawrence between Niagara Falls and Quebec; and the tickets are also valid by rail or steamer between Niagara Falls and Quebec. No extra charge for meals between Toronto and Montreal. The improved con-dition of the Grand Trunk Railway, including its equipment of new passenger cars, new locomotives, steel track, etc., now brings it prominently before the public as a first-class line, and preferable to the majority of lines between the East and West. The Grand Trunk Railway (via Gorham and the Glen House) is the only route by which parties can ascend the far-famed Mt. Washington by carriage road.

Tickets and information may be obtained at the principal ticket offices.

NEW YORK OFFICE, - - - 285 Broadway.

BOSTON OFFICE, 280 Washington St. (formerly 134).

J. STEPHENSON,
Gen'l Pass. Agt. G. T. Railway, Montreal.

ALEX. MILLOY,
Traffic Manager Richelieu & Ontario Nav. Co.
228 St. Paul St., Montreal.

(151)

CHESAPEAKE & OHIO RAILWAY,

THE NEW TRUNK LINE BETWEEN

WASHINGTON, LOUISVILLE AND CINCINNATI,

AND BETWEEN OLD POINT COMFORT, NORFOLK, NEWPORT NEWS, RICHMOND, AND

LOUISVILLE AND CINCINNATI.

SOLID TRAINS WITH PULLMAN SLEEPING CARS

BETWEEN WASHINGTON, RICHMOND, LOUISVILLE & CINCINNATI.

THE merit of this route justly entitles it to bear the appellation of "*America's Tourist Line*." There is no competitor, with the same variety of scenes, to win from it its laurels. At the extreme eastern terminus on Chesapeake Bay and looking out upon the broad Atlantic is that grand pleasure resort and sea-shore sanitarium, Hygeia Hotel, at Old Point Comfort. Hotel Warwick, Newport News, at the head of Hampton Roads, is the most elegantly furnished hotel south or west of New York. Entering the Blue Ridge and Alleghany Mountains the Mineral Springs are reached. White Sulphur, Rockbridge Alum, Warm, Hot, Healing, Sweet, Sweet Chalybeate, Red Sulphur, and numerous other resorts, embracing temples where fashion reigns— Fountains of Health presided over by Esculapius—and the pastoral homes so well suited to rest and quiet. The scenery of the entire route is grand and ever-changing from pleasant valleys to gigantic cliffs and narrow cañons.

For full information and description of the Resorts, Routes, Rates, Connections and Sleeping Car reservations, apply to any of the following named agents:

W. G. LODWICK, Ticket Agent, ... 171 Walnut Street, Cincinnati, Ohio.
JAMES C. ERNST, General Western Agent, . . 340 W. Main St., Louisville, Kentucky.
GEORGE W. BARNEY, Ticket Agent, ... Lexington, Kentucky.
W. TALBOTT WALKE, Ticket Agent, ... Norfolk, Virginia.
TICKET AGENT, Chesapeake & Ohio Railway, ... Old Point Comfort, Virginia.
H. W. CARR, General Eastern Agent, ... 229 Broadway, New York.
J. C. DAME, General South Eastern Agent, . . 513 Pennsylvania Av., Washington, D. C.

C. W. SMITH,
General Manager.

H. W. FULLER,
General Passenger Ag'n'.

(152)

RICHMOND & DANVILLE R. R.,
AND
VIRGINIA MIDLAND RY.

THE SHORTEST AND QUICKEST ROUTE TO ALL

Virginia Springs, and Summer Resorts

IN THE

Mountains of North Carolina and North Georgia.

THREE DAILY TRAINS BETWEEN

New York, Philadelphia, Baltimore, Washington, and White Sulpher, Hot, Healing, Warm, Fauquier W. S., and other Virginia Springs.

TWO DAILY TRAINS BETWEEN

EASTERN CITIES AND NEW ORLEANS, MONTGOMERY, ATLANTA

And all Summer Resorts in North Carolina and Georgia.

INCLUDING

Asheville, Warm Springs, All-Healing Springs, Flat Rock, Hendersonville, Haywood White Sulpher, Highlands, and "The Land of the Sky" in N. C., and Mt. Airy, Toccoa Falls, and Tallulah Falls in Ga.

Be sure and consult the Schedule and Rate Sheets of these Roads before determining upon your location for the summer.

FOR INFORMATION, ADDRESS,

W. A. PEARCE,
228 Washington St., Boston,

F. B. PRICE,
No. 6 N. Fourth St., Philada.,

H. MACDANIEL,
601 Pennsylvania Ave., Washington,

H. P. CLARK,
229 Broadway, New York,

JAMES HOLLINGSHEAD,
9 German Street, Baltimore,

JOHN F. McCOY,
Balt. and Potomac Depot, Wash.

M. SLAUGHTER,
General Passenger Agent.

(153)

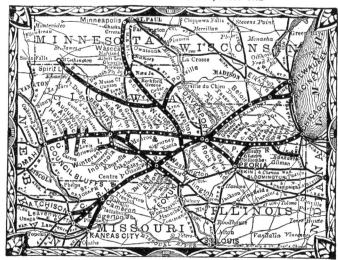

The FAVORITE ROUTE for FASHIONABLE PLEASURE TRAVEL,

VIA

THE WONDERFUL TRENTON FALLS,

Utica and Black River Railroad,

ONLY ALL RAIL ROUTE to the THOUSAND ISLANDS,

THE SHORT LINE TO

Northern New York, River St. Lawrence & Canada

TOURIST EXCURSION TICKETS

TO

Clayton, Alexandria Bay, and all Resorts among the Thousand Islands, Montreal, Quebec, White Mountains,

AND ALL

MOUNTAIN, LAKE, RIVER, AND SEASHORE RESORTS IN CANADA AND THE PROVINCES, NEW YORK STATE, AND NEW ENGLAND, BY OVER 300 DIFFERENT ROUTES.

ON SALE AT ALL PRINCIPAL TICKET OFFICES.

If you are unable to get through tickets via route you want, wait until you arrive at Utica, and purchase of H. I. Fay, agent, next the depot.

Wagner Sleeping Cars on night trains, and Parlor Cars on day trains. During summer season, fast trains, without stops, for Thousand Islands.

This route, in connection with the New York Central and Hudson River Railroad, and St. Lawrence Steamboat Company (known as the Palace Day Line), forms the New American Line for Montreal, passing all Thousand Islands and Rapids by daylight.

Send two stamps for copy of the Illustrated book, " Routes and Rates for Summer Tours." Send for a copy before deciding upon your summer trip.

J. F. MAYNARD,
General Superintendent.

THEO. BUTTERFIELD,
General Passenger Agent, Utica, N. Y.

(155)

PASSENGERS TRAVELLING BETWEEN

NEW YORK and PHILADELPHIA

WILL FIND

The "Bound Brook Route"

SHORTEST AND QUICKEST.

20 Trains Daily!   6 Trains on Sundays!

THROUGH TICKETS and BAGGAGE CHECKS to and from all PRINCIPAL POINTS.

PARLOR CARS ON EXPRESS TRAINS,
SLEEPING CARS ON NIGHT TRAINS.

3 Depots in Philadelphia, { *9th and Green Streets,*
9th and Columbia Avenue,
3d and Berks Streets.

NEW YORK STATION, FOOT OF LIBERTY STREET.

H. P. BALDWIN.
General Passenger Agent, New York.
(156)

C. G. HANCOCK,
Gen'l Pass. and Trans. Agent, Phila.

New York and New England R. R.

PRINCIPAL CITIES AND TOWNS.

Boston,   New London,   Rockville,   Danbury,
Providence,   Springfield,   S. Manchester,   Brewsters,
Worcester,   Hartford,   New Britain,   Fishkill,
Norwich,   Willimantic,   Waterbury,
And Newburg on the Hudson.

WITH CONNECTIONS FOR

New York, Philadelphia, Baltimore, & Washington,

And the SOUTH and WEST.

NORWICH LINE

BETWEEN

BOSTON and NEW YORK.

Elegant Fleet of Steamers, among them the New and Magnificient Steamer

"CITY OF WORCESTER,"

THE FINEST ON THE LONG ISLAND SOUND.

BOSTON AND PHILADELPHIA
EXPRESS LINE.

RUNNING MAGNIFICENT PULLMAN PALACE CARS,

WITHOUT CHANGE,

BETWEEN BOSTON, PHILADELPHIA, BALTIMORE, AND WASHINGTON.

A. C. KENDALL,
General Passenger Agent.

S. M. FELTON, Jr.,
General Manager.

(158)

HARPER & BRO'S TOURIST'S BOOKS.

Drake's Heart of the White Mountains. The Heart of the White Mountains. By SAMUEL ADAMS DRAKE, Author of "Nooks and Corners of the New England Coast." Illustrated by W. HAMILTON GIBSON, Author of "Pastoral Days." 4to, Illuminated Cloth, Gilt Edges, $7.50. Tourist's Edition, $3.00.

"No doubt the season will produce its regular crop of illustrated holiday gift books, but we risk little in saying that Mr. W. Hamilton Gibson's elegant volume, ' The Heart of the White Mountains, their Legend and Scenery' will remain the chosen favorite of people of good taste and artistic culture. It is printed in quarto form, and the illustrations, all from the pencil of Mr. Gibson, are beautiful in design and exquisite in execution. The letter press is by Mr. Samuel Adams Drake, and is quite worthy of the artistic part of the work. It is a superb production."—*The Sun, New York.*

"'The Heart of the White Mountains' is one of those splendid drawing-room table books which have been printed in this country only of comparatively late years. * * * This volume before us belongs to the finest class of landscape pictorials; the illustrations are little miracles of wood engraving, rendering every effect of sunlight or sunset, moonlight or gloaming, clear or foggy distances with witching illusion; there is no color, yet the feeling of the artist creates suggestions of exquisite blue and green tints. The style of the text is no doubt familiar to the readers of *Harper's Monthly*, in the pages of which both engravings and narrative first appeared. It is breezy, gossipy, instructive, and often delightfully humorous.—*New Orleans Democrat.*

"'The Heart of the White Mountains: their Legend and Scenery,' by Samuel Adams Drake, with illustrations by W. Hamilton Gibson, is one of the very few books of the year which are truly superb. As an illustration of the resources of a great American publishing house, this volume from Harper & Brothers is a matter of national pride and congratulation. It is not a merely ornamental work, lavishly adorned with fine engravings, rich binding, and other luxurious features. Its exterior is chastely elegant, not so much adorned as to destroy its character as a cover for the treasures within. * * * All but a very few of the designs are by Mr. Gibson, and this is enough to indicate their style and quality. Their engraving and printing are such as have commanded the admiration of the lovers of art in all parts of the world."—*New York Observer.*

"The Heart of the White Mountains: their Legend and Scenery. A very attractive and interesting book has Mr. Drake made out of his subject. The illustrations are excellent,—vigorous sketches, reproduced with that delicacy of execution in which the American engravers are unsurpassed."—*London Spectator.*

"This is not a dry book of travel. The weaving of incidents and legends of various places, with the description of their scenery makes this book peculiarly valuable, and seems to be a gift peculiar to Mr. Drake; not that his manner of describing these places is dull, far from it, for his vivid and almost poetical portrayals, together with the fine accompanying engravings, almost place one on the spot. He divides his book in three sections, each being the recital of a journey to the White Mountains."—*Richmond Christian Advocate.*

"'The Heart of the White Mountains,' with Drake's text and Gibson's pictures, is admired in England; the *Spectator*, after explaining to its readers where the White Mountains are, calls the book 'very attractive and interesting,' and praises the illustrations—'vigorous sketches, reproduced with that delicacy of execution in which the American engravers are unsurpassed.'"—*Springfield Republican.*

Nordhoff's California. Cloth, $1.50.

"Most of those who know anything of California are largely indebted to Mr. Nordhoff for their information. The record of his visit to the Pacific coast, some years since, was one of the most complete and reliable books that we have had on the character and resources of the state."—*N. Y. Observer.*

"People who want to know all about California will find a mine of satisfaction in the new and enlarged edition of Charles Nordhoff's California, for Health, Pleasure, and Residence. It has not a dull line in it, and gives a new idea of the resources and attractions of that state."—*Christian at Work, N. Y.*

"Harper & Brothers have issued a new and thoroughly revised edition of Charles Nordhoff's 'California for Health, Pleasure, and Residence;' the only thoroughly readable book on the subject that has yet made its appearance. It is designed for the use of both settlers and travellers, and therefore contains detailed information about the culture of the grape, the raisin, the orange, the lemon, the olive, and other semi-tropical fruits, about the methods of irrigation, the best places for colony settlements, etc."—*New York Graphic.*

"A better guide to the industries, pleasure resorts, and curiosities of the golden state, cannot be well imagined than this book on California by Mr. Nordhoff. He gives complete details regarding the culture of the vine and raisin grape, the orange, lemon, olive, and other semi-tropical fruits, colony settlements, tourists' routes, resorts for invalids, etc., etc. In short, the volume is a perfect encyclopædia of everything that is worth knowing about California, and, by its aid one can travel in imagination through a country prolific in natural curiosities and interesting developments."—*Albany Sunday Press.*

Harper & Brothers will send the above books by mail, postage prepaid, to any part of the United States, on receipt of the price.

D. O. CRANE,

Attorney-at-Law & Solicitor of Patents,

ST. CLOUD BUILDING, WASHINGTON, D. C.

PATENTS, CAVEATS, DESIGN PATENTS, TRADE MARKS, COPYRIGHTS, RE-ISSUES, Etc., SECURED.

NO FEE UNLESS SUCCESSFUL.

CORRESPONDENCE SOLICITED, PROMPT REPLIES ASSURED.

A Wheeled Vacation.

WHERE shall we go this Summer?" A question always asked, but hard to answer. The hills, the mountains, the farm, the beach, and a thousand places stare one in the face like so many enigmas, each claiming its particular advantage, and each like the guides of Rome calling "I am the best." To choose out of so varied a collection is well-nigh impossible, and he who at last settles upon one location in which to breathe the pure air and "drive dull care away," will e're his second day read of some other resort which seems to rival the place of his first choice. Many a man in trying to decide where to go has found himself like the donkey between several loads of hay, who, though dreadfully hungry, starved to death because he could not make up his mind which load was the sweetest to the palate.

Listen, and we will whisper in your ear a way to spend vacation which shall be a combination of everything and everywhere.

A bicycle tramp! a grand-free-and-easy-go-as-you-please-don't-care-for-anybody-jolly-happy-see-everything-ride from home to anywhere and back again. No horse to feed, no stable bills to pay, no weary tramp, no stay-and-get-tired-in-one-place-vacation. Alone if you will, in company is better, for the jolliest, healthiest, grandest, old time. And leave your ladies at home? Certainly not. They need the change as well as you. You on your bicycle, they on the tricycle, in a general wheel over hill and valley, out in the open air, full of life and spirits, getting health and strength, and having the best of holidays. A picnic in the woods every pleasant day, a halt for fishing in the brooks or the ocean.

What does it cost? Well, your locomotion costs nothing, your lodging a dollar or more a night, your meals what it costs you at home, unless you want to be extravagant.

(162)

Nothing is lost to the "wheelman," the nooks and corners of nature's best scenery are spread out before him. He sees everything, for he is where everything is.

What an appetite you will have; three big meals a day, and an occasional halt for milk and country gingerbread. You will leave your "steeds" untied in the road and take the most delightful strolls into the woods and over the meadows; you will sit in the cool shade of the trees beside the sparkling cascade; you will drink the cold waters of the mountain spring; you live as the birds, they with wings, you with wheels. A stop-over when you please, a day or a week, where you find a spot which will repay a larger inspection. Rest where you will, travel where you wish, and traveling will rest you, for you need the exercise.

Ah, what pleasant evenings in the kitchens of old-fashioned farm-houses, listening to the tales of bygone days; or in the parlors of the fashionable hotel, the envy of all the guests. And then to come home and back to business, flesh all aglow, blood bounding, head clear, stomach clamorous, and to sleep as you did in childhood. Dyspepsia cured if you had it, weak limbs made strong, in fact the same man renovated and repaired.

But where shall we get a machine? Why haven't you seen the "man on a wheel," which has been a "landmark" in every paper, pointing the reader to the pioneer bicycle house of America, THE POPE MANUFACTURING COMPANY of Boston, Mass., the makers of the celebrated COLUMBIA BICYCLES and TRICYCLES, the machines which have stood the test of years of wheeling over bad roads and good, by boys, men, ladies, and experts, until "Uncle Sam" has again stood upon the Capitol, waved his hat and shouted as he did over our successes in watch, piano, and sewing machine making, "AMERICA'S AHEAD." If you live near their headquarters at 597 Washington Street, Boston, step in and see the PERFECT MACHINE. If you do not, nor near one of their hundreds of agencies, then send a 3 cent stamp and receive an ILLUSTRATED CATALOGUE, which will tell you in print all about these wonderful machines.

Good as were the STANDARD and EXPERT COLUMBIAS, the improvements of this year have made these "perfect machines more perfect."

The addition of the COLUMBIA TRICYCLE to their "herd of steeds," gives the "missing link," which has separated husband and wife. To-day one can find in their warerooms "wheels" for the whole family; father, mother, boy, and girl, and even grandfather can ride the new tricycle.

> Tell us not in mournful numbers,
> Life is but a bitter dream;
> For I've wakened from my slumbers,
> And I feel so cheap and mean.
> Life is real, life is earnest;
> Life is as we do and feel,
> Happy is the life that's furnished
> Every rider of the "wheel."

COYNER'S
White and Black Sulphur Springs.

THIS WELL-KNOWN WATERING-PLACE,

Situated in Botetourt County, Virginia, on the line of, and in full view of the Norfolk & Western Railroad (only about five miles from the growing City of Roanoke), is

OPEN FOR THE RECEPTION OF VISITORS.

Since the last season NEW PORCHES have been erected to the Hotel, and the Cottages put in thorough repair.

Persons leaving BALTIMORE, WASHINGTON, RICHMOND, NORFOLK, and PETERSBURG, will arrive the same evening at the Springs; those coming from the South and West reach the Springs in about six hours from Bristol.

Visitors desiring to stop, by informing the conductor when they strike the Norfolk & Western Railroad of the fact, will be landed at the platform IMMEDIATELY OPPOSITE TO THE SPRINGS.

NO STAGING.

The Hotel being about two hundred yards from the platform makes it a VERY DESIRABLE RESTING-PLACE for persons from the South going North or returning home. Desiring to make this place a resort for families, where they can enjoy HOME COMFORTS, no trouble nor expense will be spared to render it PLEASANT AND AGREEABLE TO GUESTS.

THERE ARE FIVE SULPHUR SPRINGS, the medical qualities of which are so generally and favorably known that it is deemed unnecessary to speak of their virtues.

BOARD: Per Day, $2.00; per Week, $10.00; per Month of 4 Weeks, from $25.00 to $35.00, according to location and accommodations.

Post Office, Bonsack's, Roanoke County, Virginia. **WM. H. FRY, Gen'l Manager.**

Mount Washington House,
PARK AVENUE, BETHLEHEM, N. H.
In full view of the MOUNT WASHINGTON RANGE

C. L. BARTLETT, Proprietor.

Delightfully situated a few rods from the main street, near the new railroad station, and opposite the site of the proposed park, with the peaks of twenty-five mountains visible from its front piazza, while Mount Mansfield and the Green Mountains are observable from the back of the house. Light, airy rooms, single or *en suite* for families. Arrangements for heating the rooms of guests remaining late in the season. House especially desirable for sufferers from hay fever owing to its location.

Farms in connection. Horses and carriages, with attentive and experienced drivers.

NEW UNION HOTEL.
Hayth's Hotel and Western Hotel Combined.
FINCASTLE, VA.

Situated at Court House, in a village of 1000 population, eight miles from Jackson, on Richmond and Alleghany R. R., and six miles from Troutville, on the Shenandoah Valley R. R. Two double daily mails. Beautifully situated in a healthful locality, 1200 feet above the level of the sea. Trout and Black Bass in abundance, game also plenty in the adjoining forest. Every attraction for invalids and pleasure-seekers. Several good mineral springs near at hand.

Accommodations first-class, and rates always low.

For terms and full particulars, address,

WM. B. HAYTH, Proprietor,
FINCASTLE, VA.

Grand Central Hotel,
BAR HARBOR, ME.

This favorably located Hotel, at the great summer resort of the country, is first-class in all its appointments. Write for pamphlets and circulars to

R. HAMOR & SONS,
Mt. Desert Island, Bar Harbor, Me.

(164)

Hygeia Hotel, Old Point Comfort, Va.

Harrison Phoebus, Proprietor.

Situated on Hampton Roads 100 yards from Fortress Monroe. Accommodations for 1000 guests, and open all the year. Surroundings unsurpassed; appointments, table and service unexcelled. Boating, fishing, and driving specially attractive, and the surf bathing, which is good from May until November, the finest on the Atlantic seaboard. Terms less for equal accommodations than any health or pleasure resort in the country. Climate free from malaria and for insomnia truly wonderful in its soporific effects.

THE LARGEST SEA-SHORE AND MOUNTAIN RESORTS OF THE SOUTH UNITED.

The White Sulphur Springs,
GREENBRIER CO., W. VA.

The largest and most celebrated of all the resorts of the South, the most powerful in the therapeutic effects of its waters, the richest in natural scenic attractions, and the most favored in climate, temperature, and location, is the White Sulphur Springs, West Virginia. So widespread is the reputation of " The White " it has been called the Baden-Baden of America. Situated in the famous " Spring region" of the Virginias, on the Chesapeake and Ohio Railway, which winds through beautiful valleys, beside picturesque streams, gigantic cliffs, and narrow canons, alike accessible from the great cities of the Ohio and Mississippi valleys, the Gulf region, and those along the Atlantic seaboard, its central position invites the patronage of widely separated sections, and visitors from the North, South, East, and West come hither to enjoy the benefit of its waters, forming a society as pleasant as it is diversified. For the next five years the White Sulphur Springs Hotel, Cottages, and Restaurant will be under the management and control of the proprietor of the Hygeia Hotel, and no effort will be spared, or expense avoided, to make the combination of these two great resorts a perfect and lasting success.

Pamphlets describing the Hygienic advantages of either place will be furnished on application.

H. PHOEBUS, Proprietor and Lessee.

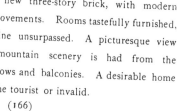

PENNSYLVANIA RAILROAD.

THE FAMOUS

"NEW YORK and CHICAGO LIMITED.

RUNS EVERY DAY IN THE YEAR BETWEEN

NEW YORK		PITTSBURGH.
PHILADELPHIA		CINCINNATI.
BALTIMORE	AND	FT. WAYNE.
WASHINGTON		CHICAGO.

THE LIMITED IS COMPOSED EXCLUSIVELY OF

Drawing Room, Dining Room, Smoking Room, and Sleeping Cars.

Meals are served in the Dining Cars at the uniform rate of $1.00.

THROUGH CAR SERVICE.

Going Westward.

One Sleeping Car Boston to Philadelphia.

One Parlor Car New York to Philadelphia.

Two Palace Sleeping Cars New York to Chicago.

Palace Sleeping Car New York to Cincinnati.

Palace Sleeping Car Washington to Chicago.

One Dining Car New York to Pittsburgh and Fort Wayne to Chicago.

One Smoking Car New York to Chicago.

Going Eastward.

Two Palace Sleeping Cars Chicago to New York.

Palace Sleeping Car Chicago to Washington.

Palace Sleeping Car Cincinnati to New York.

One Dining Car Chicago to Fort Wayne and Pittsburgh to New York.

One Smoking Car Chicago to New York

One Parlor Car Philadelphia to New York.

One Sleeping Car Philadelphia to Boston.

CHAS. E. PUGH, *General Manager.*   **J. R. WOOD,** *Gen'l Pass'r Agent.*

(167)

Two years from the Sod. No Stumps to dig out. No Stones to remove.

Real Estate as an Investment.

As interest on money is so low as to bring in but little to those possessed of a few thousand dollars, inquiry is daily made as how best to invest in order to secure sure and permanent returns.   Those in whom speculation is rife, seek to amass a fortune in a short time by dealing in delusive stocks and bonds, only to find in the end that their risk was an unsound one, and that their little fortune instead of doubling has entirely vanished.   Others, more matter of fact, are seeking investments in Real Estate in the West, where if a fortune is not made so speedily, it is sure and certain, and a heritage is assured to their children.   No Government has been more prodigal of her public lands than has ours—giving to every citizen who is desirous of obtaining it one hundred and sixty acres.   But while she has been thus kind to individual citizens, she has been more than prodigal to corporations, giving to them millions upon millions of acres of her best lands, to be parcelled out by them to honest citizens at fabulous prices.   Outside of lands thus granted to corporations, the public domain is fast being taken up by settlers, and a few years will see all our choicest lands entered.   Many persons in the East are desirous of securing lands, but their business will not permit them to go West and enter lands under the Homestead or Pre-emption laws.   There is a species of land scrip known as "Soldier's Additional Homestead" that can be located on any public lands subject to homestead or pre-emption entry, requiring no residence or cultivation, and parties can by the use of such scrip enter public land and receive title to the same without leaving their business or homes, or otherwise neglecting their affairs.

There is no limit to the amount one person can locate.   It comes in forty, eighty, and one hundred and twenty acre pieces, and is within the reach of all possessed of a little means.   There are but a few thousand acres of this scrip remaining unlocated, a majority of which I own.

For full particulars regarding location of same, price per acre, etc., etc., address,

CPSIA information can be obtained
at www.ICGtesting.com
Printed in the USA
BVHW091745021118
531990BV00019B/840/P

9 781330 230800